Observing
theDeepSky

an astronomer's companion

Observing
theDeepSky

an astronomer's companion

Darren Bushnall

The Crowood Press

First published in 2005 by
The Crowood Press Ltd
Ramsbury, Marlborough
Wiltshire SN8 2HR

www.crowood.com

British Library Cataloguing-in-Publication Data
A catalogue record for this book is available from the British
Library.

ISBN 1 86126 785 1

Illustration credits: Unless otherwise credited, all photographs
and line drawings are by the author.

Frontispiece: A barred spiral galaxy.

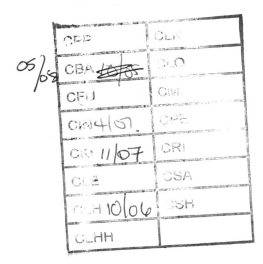

Typeset by Focus Publishing, 11a St Botolph's Road, Sevenoaks,
Kent TN13 3AJ

Printed and bound in Great Britain by Biddles Ltd, King's Lynn

Contents

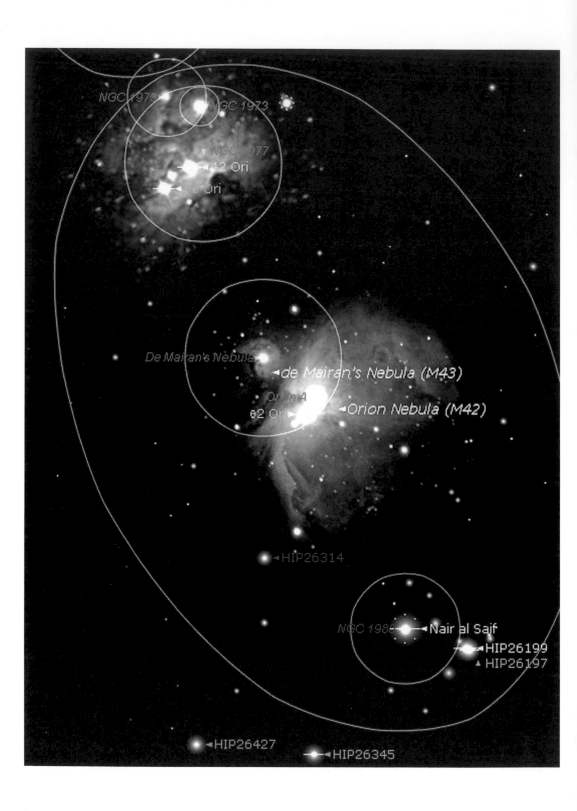

Introduction

I have always had a fascination with the night sky. I vividly remember looking at the stars way back in 1978, as a youngster of ten years. I owned no optical equipment – it was the unaided eye or nothing – but I was still fascinated, wondering what were the strange fuzzy blobs that I could see dotted here and there. I knew what the Milky Way was, but that was about it. In those days light pollution was not so much of a problem as it is now. The nights were dark and transparent, even in my home town of Hartlepool in the UK, and I could see a surprising number of objects with the naked eye.

A little later I was given my first optical instrument, a pair of 10 × 50 binoculars. To my inexperienced eyes the universe suddenly opened up: those fuzzy blobs I had previously seen only with the unaided eye were now transformed – some appearing as featureless, ghostly nebulosity, others revealed as clusters of stars – and the Milky Way exploded into a myriad stellar points. I knew how Galileo must have felt!

By the time I acquired my first telescope – a small 60mm (2¼in) refractor of dubious quality mounted on a rickety wooden tripod which seemed to shake before you got anywhere near the eyepiece – I was totally hooked. Even this small instrument with its limited light grasp and magnification could resolve most of the brighter clusters into their component stars, show detail in the brighter nebulae, and reveal many hundreds of galaxies, albeit only as fuzzy areas. Also, hundreds of beautiful double stars opened up to my gaze for the first time.

As the years passed, I became more proficient at using astronomical equipment and progressed to larger telescopes: a 220mm (8½in) f/6 Newtonian reflector and then a 300mm (12in) f/5 Newtonian, which showed fainter objects. I now use a 150mm (6in) refractor, which that suits my typical conditions and light pollution better than a larger reflector.

This could be the story of anyone starting to get interested in the night sky. All that is required is a healthy dose of enthusiasm and adventure to voyage among the stars from the comfort and safety of your own backyard.

This is a book for those with a fascination for the objects that lie beyond our Solar System. It explains how to progress in astronomy and deep-sky observing: choosing a pair of binoculars, purchasing your first telescope and deciding what other equipment you may need. There are chapters on star charts and astronomical software, techniques for tracking down objects, recording what you observe, combating light pollution, and telescope care and maintenance. At the end are extensive observing lists giving details of the northern and southern hemispheres' best deep-sky objects and double-star systems; a glossary and bibliography are also included.

Your journey starts right here.

Darren Bushnall
November 2004

1 The Eye and Binoculars

The Eye

The human eye is an organ that gathers and focuses light, and it works by the same principle as an astronomical refracting telescope: light entering the front of the eye, through the cornea and pupil, is refracted through the lens. The amount of light that enters the eye through the iris is controlled by tiny muscles expanding or contracting to adjust the aperture. As light enters the iris, the eye's lens adjusts the focal length of the eye itself, so that the light is focused on the area at the back of the eye, called the retina.

When operating in low light levels, as during observing, the muscles around the iris contract, opening it to its maximum size – usually between 5 and 7mm (about ¼in), depending on the age of the individual – enabling as much light as possible to fall on the light-sensitive retina. Figure 1.1 shows the difference between the size of the pupil under daytime light levels and when dark-adapted.

When the eye is kept in darkness away from any light, chemical reactions start to take place within the photoreceptive cells of the retina. This takes about 30 minutes to complete and is called dark adaptation (or 'dark adaption'), an important process when preparing for a night's observing. It is essential that the observer remains shielded from any lights, as once the eye is dark-adapted it takes only a fraction of a second of exposure to any bright light to totally destroy what has taken half an hour to achieve.

Cones and Rods

Inside the retina are photoreceptive cells known as rods and cones, so called because of their shape. These cells contain special pigments that are light-sensitive. Cones are clustered towards the centre of the retina, and are responsible for colour and detail vision, while rods, located to the sides of the retina, can detect only differences between light and dark but are much more sensitive to light than

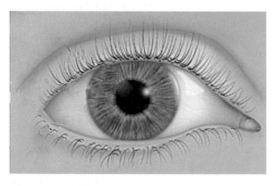 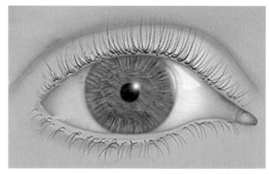

Fig. 1.1 *The eye in its normal state (left) and in a dark-adapted state (right).*

Fig. 1.2 The Milky Way from Cygnus to Sagittarius, seen from a dark observing site.

the cones. Rods are not sensitive to colour; we rely on the cones for colour detection and the rods for low-level light, this is why it is difficult to perceive colour in dark conditions.

When you look directly at an object, the light from it falls on cones, which are less sensitive to faint objects – for example, a faint galaxy. The object is difficult to see clearly, but if you look at it slightly to one side its visibility improves because the light now falls on the more light-sensitive rods. This technique, very useful to observers, is called averted vision.

What Can You See with the Unaided Eye?

The unaided eye is capable of finding over a hundred deep-sky objects brighter than 6th magnitude, which should be achievable from a dark observing site under good observing conditions. Although few look particularly impressive, it can be interesting to push the eye to its limits to detect them.

From a dark site the most noticeable 'object' is the Milky Way stretching overhead, and it is particularly impressive in the region from Cygnus to Sagittarius (Figure 1.2). Many deep-sky objects are scattered along its length. Southern-hemisphere observers can see the Milky Way's companion galaxies, the Magellanic Clouds, as large misty areas peppered with intriguing internal glows.

The Milky Way and the Magellanic Clouds are not the only galactic systems visible to the unaided eye. The Andromeda Galaxy (M31), a spiral galaxy larger than our own and lying 2.2 million light years away, is visible as a moderately bright fuzzy 'star', and keen-sighted observers under dark skies may see M33 in Triangulum, another spiral galaxy slightly more distant than M31.

A number of open star clusters can be seen with the naked eye. Some are so close and thus relatively large in our skies that they are

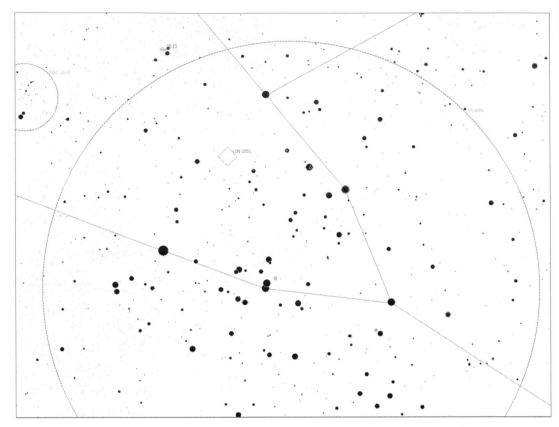

Fig. 1.3 The Hyades, a bright naked-eye cluster. (© Greg Crinklaw, SkyTools 2.)

unmistakable, for example the beautiful Pleiades (M45) in the constellation Taurus. This cluster is also known as the Seven Sisters because average-sighted people can see seven stars in the cluster; keen-sighted observers in good conditions have reported thirteen stars or more.

Just a few degrees south of the Pleiades lies another, bigger star cluster called the Hyades, spanning a diameter of ten full Moons (Figure 1.3). The bright ruddy star Aldebaran dominates this cluster but is in fact a foreground object, halfway between the Hyades and us. The Hyades lie 150 light years away; the Pleiades 380 light years.

With more distant clusters it is harder to see individual stars. A cluster may be partially resolvable, appearing as a misty glow with some stars just on the verge of being distinguished, giving it a slightly grainy look; the Beehive Cluster (M44) in Cancer, at a distance of 580 light years, is a good example. Other clusters look like misty patches of light, their stars too faint or too close together for the eye to resolve without optical aid. Examples are the Double Cluster (NGC 869/884) in Perseus, a pair of clusters both over 7000 light years away, which and appear as two circular glows almost touching.

Other deep-sky objects visible to the naked eye are nebulae, huge areas of glowing gas and dust perhaps hundreds of light years in diameter and several thousand light years away. From a light-polluted site, the best naked-eye example is the Orion Nebula (M42), visible as a fuzzy 'star' in Orion's sword (Figure 1.4).

Under dark skies the Milky Way has many areas of interest for the naked-eye observer. The North America Nebula (NGC 7000) – so called because of its uncanny resemblance to the North American continent, complete with a 'Gulf of Mexico' – is a large diffuse glow a few degrees from Deneb. Following the Milky Way down to Sagittarius will reveal other bright nebulae such as the Lagoon Nebula (M8), the Eagle Nebula (M16) and the Omega Nebula (M17).

Binoculars

Binoculars are the next step from naked-eye observing. They are highly versatile and portable, give a stereoscopic view, and have a much wider field of view than an astronomical telescope. Also, most households will probably already have a pair.

The simplest type of binocular is the opera glass, consisting of a single objective and eye lens, the same design used in Galileo's telescope. This design suffers from severe chromatic aberration (false colour) and a very narrow field of view, which makes it completely unsuitable for astronomical observing.

Modern binoculars differ from the simple opera glass by having, instead of a single objective lens, a doublet – two lenses mounted next to each other – which eliminates the false colour inherent in the single-lens Galilean design. They also have sets of prisms within the body that fold the light path, resulting in a higher-quality yet more compact design than the opera glass. There are two main types – roof-prism and Porro-prism binoculars

Roof-prism binoculars are very compact instruments, small enough to fit in the pocket and often advertised as 'travel binoculars'. They are useful for scanning the night sky to supplement observing with a telescope, though the objective lenses in this type of binocular are rather small for serious astronomy. In this design each binocular has two prisms close together, which give more of a direct light path

Fig. 1.4 A naked-eye view of the Orion Nebula.

than in other designs and keeps the size and weight to a minimum (Figure 1.5).

Porro-prism binoculars are the more familiar type, sometimes referred to as the German U-boat design, and use triangular (Porro) prisms slightly offset (Figure 1.6) to give better optical performance than the roof variety, but are more bulky.

Exit pupil

To decide whether a pair of binoculars is suitable for observing you need to take account of the instrument's exit pupil – the diameter of the beam of light as it exits the binocular eye lenses. Binoculars have numbers marked on them, often near the manufacturer's name, in the form

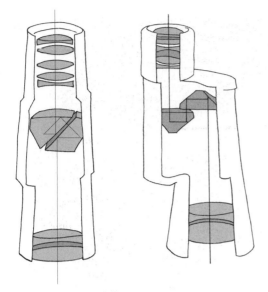

Fig. 1.5 (left) Cutaway of a pair of roof-prism binoculars.

Fig. 1.6 (right) Cutaway of a pair of Porro-prism binoculars.

8 × 25, 7 × 50 or 10 × 50, for example. The first number is the magnification, and the second is the diameter of the objectives in millimetres (Figure 1.7).

To find the exit pupil, divide the aperture (the diameter of the front lens in millimetres) by the magnification; the result is the exit pupil in millimetres. Binoculars suitable for astronomy should have an exit pupil of 5mm or larger; the larger the exit pupil, the brighter the image. Any exit pupil smaller than 5mm will give an image too dim for worthwhile astronomical use.

A pair of 8 × 25 binoculars, for example, have an exit pupil of 3.1mm, too small to be of any use in astronomy (they are more for daytime activities); 7 × 50s have an exit pupil of 7mm, and 10 × 50 binoculars have an exit pupil of 5mm, either of which is suitable for astronomical use.

Coatings

Most binocular optics are coated with at least a single layer of magnesium fluoride, which gives the familiar bluish tint to the objectives. Coatings increase the light transmission – the percentage of light that passes through the binoculars – and reduce internal reflections, thus boosting the contrast of the image. Lower-quality instruments usually only have the exterior lens coated; more expensive binoculars have multi-coated optics, meaning that all lens surfaces – exterior and interior, including the prisms – are coated, giving around 95 per cent light transmission; the higher the light transmission, the brighter the image.

Using Binoculars

A pair of 7 × 50 binoculars is not that heavy, but even after a few minutes of use your arms will ache and the image will be impossible to keep steady. Even your heartbeat will make the image jump around alarmingly, even at a magnification as small as ×7. The problem worsens if you use binoculars with a higher magnification, such as 10 × 50s: the higher power will magnify any tremors in your hands and arms even more, making worthwhile observations impossible.

To overcome this, attach the binoculars to a sturdy tripod. Not only will you then have a steady image, you will also have your hands

Fig. 1.7 Close-up of a pair of binoculars, showing the magnification and the diameter of the objectives in millimetres, here 8 × 40.

Fig. 1.8 An adapter for attaching binoculars to a tripod.

Fig. 1.9 A simple box for mounting binoculars on a tripod.

free to do other things such as sketching. Some binoculars have a threaded screw to take a special adapter which can be attached to a tripod (Figure 1.8). If there is no such means of securing the binoculars to a tripod, then a simple box constructed from plywood is easy to make (Figure 1.9). This box has a hole drilled in the base where a short bolt threads through the tripod head and is secured with a nut. The binoculars simply rest in V-shaped cut-outs and are secured by strong elastic bands.

Big Binoculars

Binoculars such as 7 × 50s and 10 × 50s are fine for general observing, and have the added bonus of being useful for daytime use, but by observing a with pair of large astronomical binoculars you will appreciate the beauty of the night sky even more. The term 'big binoculars' generally means objective lenses of 60mm or larger and a magnification of ×10 upwards. Typical sizes are 12 × 60, 11 × 70 and 16 × 80 – and they also come even larger than that (Figure 1.10).

Fig. 1.10 A pair of 16 × 80 binoculars mounted on a tripod via an adapter.

The main criterion for choosing a pair of big binoculars is the same as for choosing a pair of 7 × 50s: any instrument with an exit

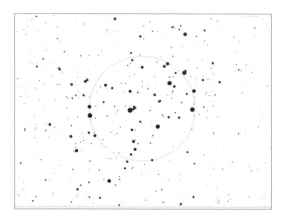 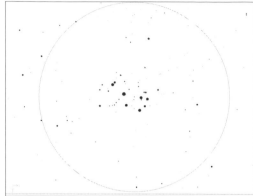

Fig. 1.11 (a) A typical low-power (×40) telescopic field is too small for large open clusters. Here, only the central members of the Pleiades are visible in the field of view. (b) Binoculars give a more pleasing view. (© Greg Crinklaw, SkyTools 2.)

pupil smaller than 5mm is not suitable for astronomy. In addition, big binoculars are weighty beasts, and with their higher magnification a secure mount is absolutely essential.

What Can You See with Binoculars?

Deep-sky objects with a large apparent diameter are better viewed with binoculars than with a telescope. A typical field of view of a telescope used at low power is 1° across, which is two Moon diameters, whereas a pair of standard 10 × 50 and 7 × 50 binoculars have typical fields of 5° to 7° – that's 10–14 Moon diameters, a huge difference compared with the telescope.

Objects that are better suited for binoculars are large open clusters such as the Pleiades, in Taurus. This cluster spans 1° of sky, and its overall effect is lost when viewed through a telescope (Figure 1.11a), but binoculars with their lower magnification and much wider field of view give a more aesthetically pleasing view (Figure 1.11b). Also in Taurus is the huge open cluster the Hyades. This is about 5° in diameter, far too large for even the lowest-power, widest-field telescope, but beautifully framed in binoculars.

A good pair of binoculars will reveal a number of galaxies. The Andromeda Galaxy, visible with the unaided eye as a tiny fuzzy patch, becomes a bright, elongated streak of light in binoculars. In the nearby constellation of Triangulum, the large face-on spiral galaxy M33 (also just visible to the naked eye) is seen as a circular fuzzy patch. Many more galaxies are within the reach of a typical pair of 10 × 50 binoculars in good observing conditions.

Binoculars will show larger nebulae well. The magnificent Orion Nebula viewed through securely mounted 10 × 50 binoculars is a complex patch of light with a few stars embedded in its glow (Figure 1.12). Smaller nebulae are also visible, such as the Dumbbell Nebula (M27) in Vulpecula, seen as a small bright disk of light. Scanning the summer Milky Way from Cygnus down to Sagittarius will show countless objects, and the Milky Way itself resolves into uncountable stellar points.

Observers in the southern hemisphere can also use binoculars to good advantage. From southerly latitudes the Milky Way is a truly magnificent sight: the galactic centre, which is located in Sagittarius, can be directly

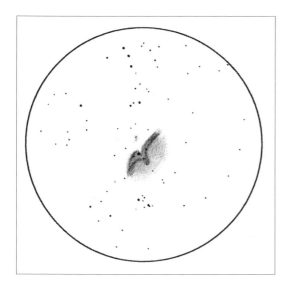

Fig. 1.12 With their larger apertures and higher magnification, astronomical binoculars reveal more detail in deep-sky objects.

overhead. Many naked-eye glows are resolved into areas of bright nebulosity, a good example being the Lagoon Nebula (M8) in Sagittarius, which is visible as a complex, irregular patch of light peppered with stars; also in the same low-power binocular field is the Trifid Nebula (M20).

There are many star clusters for which the wide field of binoculars is perfectly suited. An obvious candidate is the naked-eye cluster M7 in Scorpius, which is resolved into stars even in small binoculars. This cluster is well over 1° across – rather too large for a telescope, in which the effect of a cluster is lost, but since a modest pair of binoculars will have a field of at least 5°, the cluster is framed beautifully.

Scanning the southern Milky Way with binoculars will reveal many areas that seem devoid of stars. These areas are dark nebulae – vast regions of opaque dust and gas, which if situated in front of something bright give the effect of holes in space. Many dark nebulae are large and therefore only suitable for instruments that give wide-field views. A good example is the Coalsack in Crux, visible as a large irregular patch against the bright Milky Way starfields.

Many galaxies are visible in binoculars but aren't at all spectacular. However, observers in far southern latitudes can see the Milky Way's two satellite galaxies, the Magellanic Clouds. These two irregular galaxies are far too large to be viewed in their entirety with any telescope, but binoculars have fields wide enough to contain them, giving a much more pleasing view. In addition there are dozens of objects in the Magellanic Clouds themselves, such as the Tarantula Nebula (NGC 2070) – the largest known emission nebula in the heavens, and well seen in binoculars.

With their larger aperture and higher magnification, typical 16 × 80s will begin to show detail in many objects: open clusters that are merely patches of light as viewed in smaller instruments resolve into individual stars, more structure is visible in nebulosity and many more galaxies pop into view.

2 Telescopes and Mountings

Binoculars are ideal for starting out, but the more serious stargazer will want to upgrade to an astronomical telescope. In today's high-tech world the amateur astronomer is spoilt for choice. Go to any major astronomical equipment retailer and you will find every conceivable type of telescope on display, from the smallest spotting scopes to large Dobsonian 'light bucket' reflectors, with a price range to suit every pocket. The type of telescope you buy will depend not only on what you can afford, but also on where you live, where you plan to use the telescope, and what type of observing you intend to do.

Types of Telescope
Refractors
Many amateurs will probably start with a small refracting telescope with an aperture of 60–80mm (2¼–3in), a little over a metre long, on a simple altazimuth or equatorial mounting. Telescopes of this size and of a reputable brand are often described as achromatic ('without colour') or achromats, indicating that they have an objective lens which is a doublet, made from two different types of glass, usually crown glass and dense flint glass (Figure 2.1).

There are many good small refractors by reputable astronomical manufacturers, any of which would make an excellent introduction to telescopic astronomy. But beware – there are plenty of poor ones too. In particular, avoid cheap department-store telescopes, which usually have an aperture of 50mm (2in) and boast massive magnifications on their brightly coloured boxes. These telescopes sometimes have an objective made from plastic rather than glass! Also, instead of being a doublet the objective is sometimes a single lens, as in the opera glass, which causes massive amounts of false colour plus terrible distortions around the edges of the telescope's field of view. To try to overcome this problem, manufacturers of these telescopes put an aperture stop just behind the front lens. This is a piece of opaque plastic with a hole in it that is even smaller than the advertised aperture, to cut down on the aberrations. The result is an even dimmer image, making the instrument totally useless for astronomy.

Even a good achromatic doublet suffers from some false colour, especially in telescopes of shorter focal ratio. A telescope's focal ratio is its focal length in millimetres, which is usually marked on the telescope tube or near the

Fig. 2.1 *The light path through a typical achromatic refractor.*

Fig. 2.2 A large 150mm (6in) refractor.

focuser, divided by its aperture. For example, a 150mm (6in) telescope with a focal length of 1200mm will have a focal ratio of 8 (written as f/8). The false colour evident in achromatic telescopes is caused by light rays passing through the objective being split into their component colours; each colour is brought to a slightly different focus, and as a result the image of any bright object is surrounded by a bluish halo, which distracts the eye and obliterates detail.

Refractors with a focal ratio of 10 and below – which means the majority of commercially available instruments – tend to suffer more from false colour. In refractors with focal ratios of 10 to 15, false colour is progressively minimized, but they have long and unwieldy tubes: the longer the focal ratio, the longer the physical length of the instrument.

While small telescopes give pleasing views, much better is the view through a refractor with an objective from 100 to 150mm (4–6 in) (Figure 2.2); such instruments give razor-sharp images, dark background fields and very good contrast. Refractors in this size range give better views than similar-sized reflectors and catadioptrics because it is far easier to achieve a precise optical figure on a lens than on a mirror; refractors have no central obstruction, as do the other designs; and a refractor has a sealed tube, which greatly reduces the temperature variations within the tube that create the air movements that may spoil the view in open-tube designs.

Refractors in the 100–150mm range vary considerably in price and performance. Like smaller refractors, larger instruments at the lower end of the price range have achromatic doublet objectives (usually with the two elements spaced, rather than in contact) to bring the red and blue components of light to a close focus, but you do still get some false colour. Some of the more expensive instruments have an objective doublet coated with calcium fluorite, which significantly reduces the chromatic aberration.

Finally, there are apochromatic refractors, or apochromats, whose objectives consist of three lenses rather than two. The objectives in 'apo' systems use three different types of glass, which gives images of the purist optical quality – the blue haloes around bright objects caused by chromatic aberration are eliminated even in instruments of very short focal length. But apochromatic refractors are very expensive.

Refractors do have their drawbacks. The main one is the physical size of the instrument (as mentioned above): a 100mm (4in) f/15 refractor is over 1.5m (5ft) long because of the straight-through light path. In addition, because the eyepiece is at the lower end of the instrument the telescope has to be mounted high above the ground if the eyepiece is to be at a convenient height for observing objects near the zenith. The length of refractors also makes them difficult to mount steadily – they tend to sway in the slightest breeze if the mounting is not sturdy enough.

Another drawback of the refractor is its limited light grasp, especially in smaller apertures. The great majority of deep-sky objects are faint and require as much light-gathering power as possible. The ever-popular 100mm (4in) refractor is quite limited in what it can see compared with other telescope designs; it will show many deep-sky objects, but no significant detail in them.

For observing double stars this limited light grasp is not such an issue, for even a small aperture such as 100mm is capable of resolving over 10,000 double stars. In addition, because refractors have an unobstructed light path the image is purer and free from diffraction spikes, which may hide fainter companion stars (*see* Chapter 12), that you get with instruments that have a central obstruction, such as the Newtonian.

Reflectors

The reflecting telescope (Figure 2.3) is usually the instrument of choice for the deep-sky observer, because deep-sky objects are often very faint and thus require as large an aperture as possible, and, aperture for aperture, reflectors are cheaper than any other type of telescope. For the price of a good 100mm (4in) refractor, you could buy a 250mm (10in) reflector capable of showing many more

Fig. 2.3 A 220mm (8½in) Newtonian reflector.

objects. Also, you have much more choice when buying a reflecting telescope: there is every size of instrument for the amateur from small 100mm scopes to monster reflectors over 2 metres high with apertures of 750mm (30in).

Reflectors differ from refractors in that they use a curved mirror to focus the light rays (Figure 2.4). Light strikes a paraboloidal mirror situated at the lower end of the tube, reflects back up the tube onto a smaller elliptical mirror, called the secondary mirror, diagonal or flat,

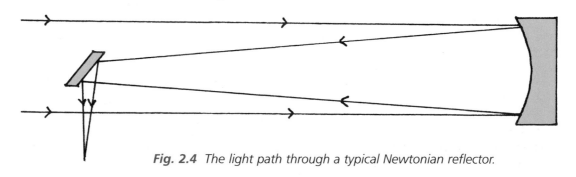

Fig. 2.4 The light path through a typical Newtonian reflector.

set at a 45° angle. From there the light is reflected to the side of the tube, where it forms an image which is then magnified by the eyepiece.

Since light rays in a Newtonian system merely reflect off the mirrors and do not pass through an objective lens, as they do in the refractor, chromatic aberration is not an issue: images in the reflector are not surrounded by coloured fringes. Of course, any telescope that uses a mirror could still suffer from other forms of aberration such as astigmatism, coma and spherical aberration. However, most aberrations are usually found in instruments with a focal ratio of f/6 and lower, because for these it is harder to manufacture a perfect paraboloidal figure for the primary mirror.

Reflectors make great portable instruments: a typical 100mm (4in) reflector is only about 1.2m (4ft) long, and because the eyepiece is situated at the top of the telescope's tube you have a very comfortable viewing position.

Reflecting telescopes offer the best value for deep-sky observing, as large apertures are available for a relatively modest outlay and allow more light to be gathered from faint objects than would a refractor of comparable cost. But the small secondary mirror in the Newtonian's optical path does block some of the incoming light, reducing contrast. In addition, the tube of a reflecting telescope is open at the top end, which can cause problems with 'tube currents'. When a telescope is taken from a warm room into cold air, warm air rises from the bottom of the tube to the top, where it cools and drops down again; this cycle can repeat for quite a while until the air inside the tube and the tube itself reach a temperature equilibrium. Tube currents make the view very poor and ill-defined, until the instrument cools down; the larger the aperture, the worse the effect.

Catadioptrics

Catadioptric telescopes combine elements of the refractor and Newtonian designs,

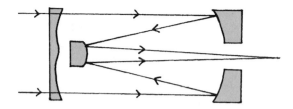

Fig. 2.5 *The light path through a Schmidt–Cassegrain telescope.*

combining mirror and lens, and because they have a folded light path they are very compact and portable.

The most popular catadioptric design is the Schmidt–Cassegrain telescope, or SCT (Figure 2.5). This telescope has a spheroidal primary mirror rather than the paraboloidal one used in the Newtonian. This spheroidal mirror would on its own produce large amounts of spherical aberration, but the catadioptric has a specially figured convex corrector lens to remove any traces of spherical aberration in the image. The corrector is mounted at the front of the tube, sealing the tube from any dust, and it holds the secondary mirror. As this type of telescope has a long focal ratio, of at least f/10, aberrations that can plague Newtonian reflectors of shorter focal ratio are not as common.

Another popular design is the Schmidt–Newtonian reflector. It works in the same way as the SCT, but with one difference: the light that has been reflected by the spheroidal primary mirror then hits a flat, as in the Newtonian reflector, and leaves the tube through the side, not through a hole in the mirror. The lower focal ratios of these telescopes make them well-suited for deep-sky observations.

The catadioptric design ranges from small spotting scopes with apertures of 90mm (3½in) to 355mm (14in) giants. Because the light path is folded three times, the catadioptric design is good for portability: a 200mm (8in) f/10 SCT is only about 0.6m (2ft) in length, and when mounted on a fork

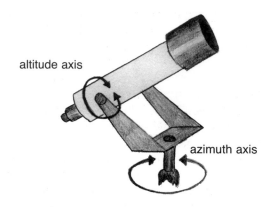

altitude axis

azimuth axis

Fig. 2.6 *The simple altazimuth mounting allows the telescope to move about two axes.*

mounting (*see below*), as catadioptrics often are, is a very stable system.

Catadioptrics make good deep-sky telescopes as they come in a range of suitable apertures, the largest of which are suitable for serious deep-sky studies. An extensive range of accessories are commercially available, and up to an aperture of 250mm (10in) they are very portable. However, the SCT's secondary mirror is located in the light path and blocks around 30 per cent of the incoming light, which does reduce contrast, and is an unavoidable feature of the design.

Telescope Mountings

A good telescope mounting holds the instrument itself firmly in place; a top-quality instrument is unusable if compromised by a shaky, unstable mounting. Ideally, the mounting should be massive as possible: larger than the telescope actually needs, to ensure vibration-free images, but this is not always practical. A large, stable mounting is a heavy piece of equipment – ideal if permanently set up in an observatory, but if you need to transport the telescope to dark skies such a mounting will be impossible to move easily. A mounting that is small and light enough to be moved will undoubtedly be slightly shaky with

the added weight of the telescope and counterweights, but that is the price of portability.

The Altazimuth Mounting

There are two types of telescope mounting. The simplest is the altazimuth (Figure 2.6), which allows telescope to move in altitude (up and down) and azimuth (left and right). This allows the observer to place the instrument anywhere and start observing as no polar alignment is required, as it is with the equatorial mounting (*see below*), but the disadvantage of this mounting is that the observer cannot track celestial objects. To follow an object across the sky, the telescope constantly needs repositioning by hand or by slow-motion cables attached to the mounting. (These are cables, usually attached to the altitude axis of the telescope mounting, which enable the observer to slowly and precisely move the instrument by hand.)

The simple altazimuth mounting is most often used for small refracting telescopes. The Dobsonian telescope (*see below*) is a popular adaptation of the Newtonian reflector. However, its inability to track objects makes it unsuitable for astrophotography, unless it is used with a Poncet platform (*see* Chapter 4).

The Equatorial Mounting

Most commercially available mountings are of the equatorial type. This type of mounting has two axes: the polar axis (right ascension), and the declination axis (Figure 2.7). The mounting is set up so that the polar axis points to the visible celestial pole, a job made easier for northern-hemisphere observers because the North Pole Star, Polaris, is just under 1° from the true north celestial pole. When the mounting is correctly aligned, moving the telescope about the polar axis to compensate for the Earth's rotation will keep an object in the field of view, a procedure which can be automated by using a clock drive on the mounting.

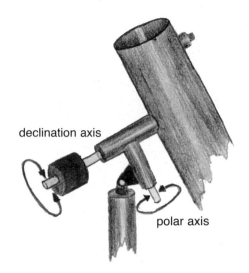

declination axis

polar axis

Fig. 2.7 The German equatorial mounting.

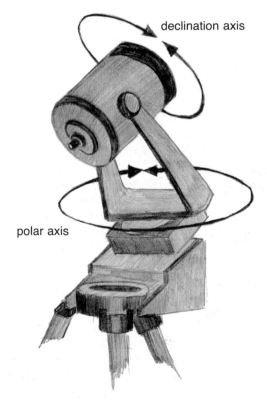

declination axis

polar axis

Fig. 2.8 A fork mounting.

By far the most common equatorial mounting is the German type. The declination axis holds the telescope, and the polar axis carries a counterweight to balance the system. Another popular mounting is the fork – as it name implies, a large fork at the upper end of the polar axis, inside which the telescope tube can swing (Figure 2.8). The advantage of the fork mounting over the German type is that the telescope is attached to both fork arms, so no cumbersome counterweight is needed, but it is suitable for only short instruments such as the popular Schmidt–Cassegrain telescope.

Which Telescope Is Best for Me?

There is no one 'best' telescope. All three types – refractor, reflector and catadioptric – have their own advantages and weaknesses, and what you decide to purchase will depend on several factors. How much are you willing to spend? Are you going to house the telescope permanently in an observatory, or will you want to be able to transport it to a dark site? What types of object do you wish to study?

For the observer who wants to observe all types of deep-sky object, but is on a budget or doesn't have room to store a large telescope, a simple 115mm (4½in) reflector on a simple equatorial mounting or a 150mm (6in) Dobsonian telescope will suffice. These telescopes are very portable, small enough to hide away in a corner and capable of showing many wonderful sights in the night sky – and they will not break the bank. Alternatively, one may purchase one of the excellent 80mm (3¼in) or 90mm (3½in) refractors available from various manufacturers. These telescopes are available in long and short focal lengths, and those of shorter focal length will easily fit on the back seat of a car.

The more serious observer may opt for a 200mm (8in) or 250mm (10in) reflector, either of which will show detail in many of the brighter objects and will be compact enough, if on a Dobsonian mounting, to be very portable – though Newtonian reflectors of over 250mm aperture can be awkward to transport. The typical Dobsonian telescope consists of a normal Newtonian perched on a wooden box-and-cradle structure (Figure 2.9). Large Dobsonians used to be quite cumbersome to transport, but the more recent 'truss' designs replace the tube with a framework of struts, and can be dismantled or assembled into a full telescope within minutes (Figure 2.10). This type of construction is usually seen on very large apertures, of say 400mm (16in) and above, though it is not uncommon to see truss-type Dobsonians as small as 300mm (12in).

If storage is a problem, a 125mm (5in) catadioptric on a small equatorial mount may be ideal. A telescope of this size is quite short yet can give superior views to a reflector of similar aperture. However, larger catadioptric telescopes are very expensive. An alternative to the catadioptric is a Schmidt–Newtonian, similar to the Newtonian reflector in appearance but with a mirror and a corrector lens supported by an optical window. These telescopes give refractor-like quality and wide-field views, and are highly portable.

The Schmidt–Cassegrain has been the amateur's favourite catadioptric for over thirty years. The first models were very basic but very usable telescopes, while today's SCTs have just about every piece of astronomical technology available integrated into the mounting. They are available in just about every size imaginable, and their folded light path makes them highly compact instruments (Figure 2.11).

The main disadvantage of the catadioptric design, which includes the SCT, is that it uses a lens at the front of the telescope held in place by an optical window, which blocks some of the incoming light, dimming the image slightly and reducing contrast. The Newtonian reflector also has this disadvantage, with its secondary mirror in the light path.

Fig. 2.9 The highly portable Dobsonian telescope.

Fig. 2.10 A truss-design Dobsonian.

Fig. 2.11 A compact yet highly versatile SCT.

What Aperture Do I Need?

When purchasing a telescope the deciding factor will be its aperture, the size of the main mirror or lens. The larger the aperture, the fainter the object you will see and with finer resolution. It may seem that, given a large enough budget, choosing the largest possible aperture is the way to go, but large telescopes are a mixed blessing. The larger the aperture, the larger and heavier the telescope. Where are you going to store that 100kg (220lb), two-metre high 400mm (16in) Newtonian reflector you so desperately want? Could you move it by yourself? A telescope that is difficult to carry around is a telescope that will probably stay indoors, so perhaps that smaller, more manageable 200mm (8in) reflector is a better choice after all (Figure 2.12).

Where you live will also determine the largest aperture that you will be able to use. Large mirrors collect more light than small ones, but a large aperture is only practical if you can observe well away from urban lighting: a large mirror will capture more light from deep space, but will also capture light from other sources like street lighting (*see* Chapter 9), which will give you poor, washed-out views. You may be able to transport the instrument to a dark site, but then you have the problem of the physical size and weight of the telescope.

Your first telescope has to be a compromise: small enough for easy portability, unless housed permanently in an observatory, but

Fig. 2.12 Smaller telescopes are more portable than larger instruments, as they break down into more manageable pieces.

with a large enough aperture for serious deep-sky study – not so large, though, to be seriously affected by local light pollution.

Buying a telescope
Where to Buy –
Local Dealer or Internet?

There are astronomical retailers located in every major country. In the United States there is a telescope dealer in nearly every large city. The United Kingdom has only a handful in the entire country, making a trip to browse, quiz the sales rep and possibly purchase your new instrument more difficult depending on where you live.

If as a first-time telescope buyer you are unsure of what type of instrument you need, actually going to an astronomical retail outlet and talking to the staff can be valuable in helping you to choose. You will be able to leave with your new purchase happy in the knowledge that you have made the right choice.

If it is not possible to visit a retailer, then the internet can be a wonderful place. Every major supplier has a website, so you can browse all the telescopes on offer, read the specs, look at prices, go to another website, compare prices –

and possibly get a better deal. This method is fine if you know exactly what you want. If not, it's better to talk to other observers or visit your local astronomical society to get as much information as possible.

If you do purchase online, make sure you know what you are buying. Different telescopes can have similar model numbers, and you don't want to open the box containing your shiny new 200mm reflector only to find that you have actually ordered a 100mm because the last few digits of the order code were different! Also, read any small print on the website regarding return policies and warranties – though with any reputable retailer this should not be a problem.

Finally, if buying via mail order, check that the price includes shipping. Large, heavy instruments can come with rather a large delivery bill.

Buying Second-Hand

For those on a limited budget a used telescope may be the best choice. 'For sale' ads are placed on various internet sites and in many monthly astronomical magazines, so it is possible to pick up a real bargain. But buying second-hand does carry risks, especially from advertisers who state 'buy as seen'.

Always try to inspect the telescope before you hand over any money. Check all the moving parts – are they smooth, do any moving parts make any unusual sounds when moved? If it has a clock drive, make sure that it is fully operational and pay close attention to the gears – check that no teeth are broken or bent.

Most importantly, inspect the optics thoroughly. Look for any scratches on a refractor's objective lens – scratches scatter light and reduce definition. Check the mirrors in a Newtonian reflector: these telescopes have an aluminium coating on the mirrors that will tarnish over time, reducing their reflectivity and significantly lowering contrast and definition. If the coating is defective, try to get the price of the telescope reduced to compensate for the cost of re-aluminizing the optics.

The primary mirror of a catadioptric telescope optics is not exposed to the elements, so its coating has a considerably longer life. But do check the optical window at the front of the telescope: make sure that its coating is intact and that there is no physical damage.

If you don't have the experience to check these things yourself, it is best to take a knowledgeable person with you.

3 Eyepieces

Eyepieces are the last link in the telescope's optical chain. No matter how good your telescope is, if you use a cheap, poor-quality eyepiece the image will be dull, fuzzy and disappointing. There are many designs on the market. Their names, such as Nagler, orthoscopic and Plössl, and their technical specifications, can be confusing to the novice, so it pays to do your homework before you buy. Some eyepieces cost more than a basic telescope.

What Magnification Do I Need?

Before you decide on a set of eyepieces, you need to work out a range of magnifications suited to the type of observing you will be doing. An eyepiece doesn't actually magnify by a set amount: its power depends on the focal length of the telescope it is used with. A set of three or four eyepieces of various focal lengths, used perhaps with a Barlow lens (see Chapter 4), should cover all types of observing.

To work out the magnification given by an eyepiece, divide the focal length of the telescope in millimetres by the focal length of the eyepiece in millimetres. So, for example, a 1000mm focal length telescope used with a 25mm focal length eyepiece will yield a magnification of ×40.

To get the best out of your telescope, three or four eyepieces will be all you need: a low-power eyepiece (×30–50) for large extended objects such as sprawling open clusters or nebulae, a medium-power one (×80–120) for

hunting down galaxies and resolving more compact objects, and a high-power one (×150–250) for finding tiny planetary nebulae and resolving detail in galaxies. In addition, an ultra-high-power eyepiece, with a magnification of around ×300–400, may be useful for separating close double stars or resolving a disk on a virtually point-like planetary nebula. But observing conditions will rarely warrant such high powers: the 'seeing', as reckoned on the Antoniadi scale (see Chapter 7), is seldom ideal. To determine the highest magnification that can be used with an astronomical telescope, multiply the aperture in inches by 50. For example, the maximum magnification usable with a 150mm (6in) telescope is ×300 – but it would require exceptional seeing conditions to use such magnifications. In practice, the highest magnification a telescope will take is more like ×30 per inch of aperture.

Eyepiece Characteristics

The specifications for eyepieces include features such as barrel diameter, apparent field of view, exit pupil and eye relief. You will need to understand these to be able to choose the eyepieces suited to your needs.

Barrel Diameter

The barrel diameter is simply the diameter of the eyepiece. Small, cheap telescopes usually come with 24.5mm (0.965in) eyepieces, designed to fit in a focuser of the same diameter, and are usually of the Huygenian or

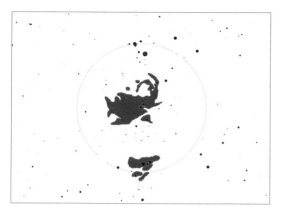

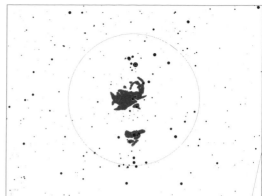

Fig. 3.1 Simulated views of the Orion Nebula, through eyepieces with a an apparent field of (a) 52° and (b) 80°.

Ramsden designs (*see below*). Most telescopes come with a 31.5mm (1¼in) focuser and eyepieces. Many different makes and designs of eyepiece, including wide-angle ones, are available to fit this standard barrel size. Some telescopes will take 50mm (2in) eyepieces, which give super-wide fields.

Apparent Field of View

Different eyepiece designs have different apparent fields of view – the angular size of the field that you will see when you look into the eyepiece. The size of the apparent field is determined by a metal ring called a field stop located inside the eyepiece. If you hold an eyepiece up to a bright light source and look into it as you would if it were in a telescope, you will see a bright circle of light: the angular diameter of this circle is the eyepiece's apparent field of view. If you use this method to compare, for example, a Kellner and an Erfle eyepiece, you will see that the circle of light is much larger in the Erfle, which has an apparent field of around 60° as opposed to the Kellner's 45°. The larger the apparent field of view, the wider and more impressive the view through the eyepiece.

Figure 3.1a shows how the Orion Nebula would appear through a standard eyepiece with a 52° apparent field and a magnification of ×40. Figure 3.1b shows how much more can be seen through an eyepiece giving the same magnification but with an 80° apparent field.

Most standard eyepieces, such as the orthoscopic and Plössl, have apparent fields of 30° to 50°, while specialist eyepieces can be as wide as 82°. But an eyepiece that has an 82° apparent field will not, of course, allow you to see an 82° diameter region of sky when used with the telescope. For the real field of view, you need to divide the apparent field of the eyepiece by the magnification it gives with your telescope. So, for example, an eyepiece with a 50° apparent field that magnifies by ×50 will have a true field of 1°.

Exit Pupil

The exit pupil is the smallest diameter of the beam of light as it leaves the eyepiece. It is usually expressed in millimetres, and is found by dividing the diameter of the of the telescope's primary lens or mirror by the magnification of the eyepiece, just as described for binoculars in Chapter 1.

Eye Relief

Eye relief is the distance you need to place your eye behind the eyepiece to see the full field of view. Generally, the higher the magnification of an eyepiece, the smaller the eye relief. Some eyepieces have an eye relief so small that you have to press your eye right up against the eyepiece to see the full field, something that spectacle wearers need to take into consideration, as the spectacles will need removing.

Coatings

All modern eyepieces are multi-coated with magnesium fluoride. Many top-quality eyepieces are fully multi-coated, meaning that all their internal surfaces are coated to prevent internal reflections and improve light transmission. This is especially important in wide-field eyepieces, which can have up to eight component lenses. In addition, look for eyepieces whose lenses have blackened edges, which improves contrast.

Aberrations

Most modern eyepiece designs give very good images, though some do suffer from optical aberrations. Simple eyepiece designs can show spherical aberration, caused by light rays passing through the centre of the eye lens and coming to focus slightly behind the rays that pass through the edge of the lens, resulting in an unsharp image.

Another common aberration is field curvature, when an object in focus at the centre of the field is out of focus away from the centre, and vice versa. Field curvature is present to some extent in most wide-angle eyepiece designs.

Cheap eyepieces often exhibit astigmatism, caused by light rays coming to focus in different parts of the lens, which can make stars look like little people or seagulls. Astigmatism is a common problem within the human eye, so one should be wary if you test an eyepiece and detect astigmatism – make sure that the fault is with the eyepiece and not your eye.

Eyepiece Designs

Choosing eyepieces can be a daunting task. There are many designs, with different numbers of lenses and lens configurations. Some are much more expensive than others, and some work better with certain types of telescope.

Standard Eyepieces

The simplest types of eyepiece are the Huygenian (Figure 3.2a) and the Ramsden (Figure 3.2b), which are often supplied with cheap telescopes. With only two lenses, the Huygenian has an apparent field of 30°–35°, which is like rather looking through a drinking straw. The Ramsden has a wider field of 35°–40°, still narrow compared with other designs. These eyepieces, despite their simplicity, work well with long-focal-length telescopes such as small refractors (which are usually over 1000mm focal length – f/10 and higher), but because of their simple lens arrangements and small apparent fields they give poor views in most of today's lower-focal-ratio telescopes. In general, the Huygenian and Ramsden eyepieces now find little use in deep-sky astronomy.

The Kellner (Figure 3.2c) is an improvement on the Huygenian and Ramsden designs, having three lenses and a larger apparent field, typically 45°. It gives bright, sharp images at low to medium powers and works best on telescopes of f/8 and larger, though it does suffer from some chromatic aberration. Most commercial small refractors, and reflecting telescopes such as the popular 115mm (4½in), are f/8 or more, so a Kellner eyepiece would be suitable for an inexpensive upgrade to replace the poor-quality eyepieces often supplied with these instruments.

A better eyepiece is the orthoscopic (Figure 3.2d), a four-element design which provides a well-corrected 45° field suitable for all general observing, especially planetary observation. Generally, the orthoscopic has good eye relief and excellent sharpness and colour correction,

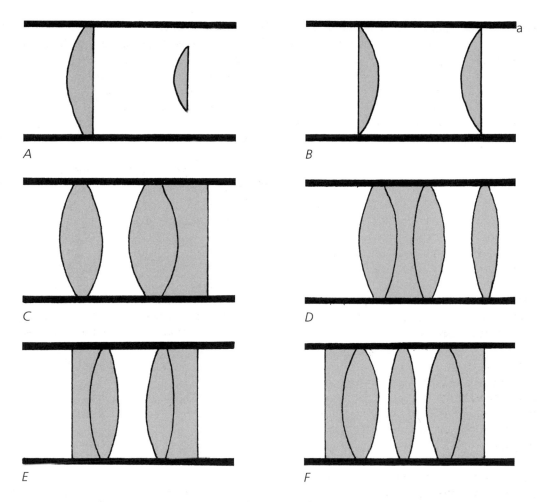

Fig. 3.2 *Types of eyepiece: (a) Huygenian, (b) Ramsden, (c) Kellner, (d) orthoscopic, (e) Plössl, (f) super-Plössl.*

but it does suffer from a little field curvature and internal reflections if viewing bright objects. The orthoscopic design works well with instruments down to f/6, which is to say with the majority of the 150–300mm (6–12in) telescopes available, so will make an excellent eyepiece upgrade for the observer on a small budget, though its 45° apparent field will not give very wide views.

The most widely used standard eyepiece in amateur astronomy is the Plössl (Figure 3.2e), four-element design. It gives a well-corrected 52° apparent field with only slight field curvature and, compared with the orthoscopic, provides comfortable eye relief – which is good news for spectacle wearers. A popular improvement is the super-Plössl, a five-element design which eliminates all field curvature, astigmatism and internal reflections (Figure 3.2f).

The Plössl is general accepted as the standard eyepiece design for deep-sky observing. It works well with telescopes of very low focal ratio, providing a pleasing wide view of the heavens, and is not particularly expensive. The super-Plössl costs a little more;

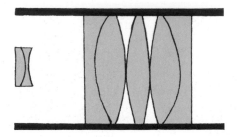

Fig. 3.3 *A lanthanum eyepiece with a built-in Barlow lens.*

it gives the same (or in some cases a slightly wider) apparent field, and because of its extra lens it works better than the standard Plössl with instruments of lower focal ratio. The wider the apparent field of view, the more lenses the eyepiece needs to correct any distortions that may originate in the outer region of the field; as the light path needs to travel through more glass, potentially scattering light before it reaches the eye, these eyepieces have every glass surface fully multi-coated to maximize the light transmission. Modern super-Plössls are designed by computer (as are pricier eyepieces) to provide the very best views through modern telescopes.

In recent years a newer eyepiece has become available, called the lanthanum. Originally manufactured by Vixen but subsequently copied by other manufacturers, these eyepieces have lenses made from glass that incorporates lanthanum, a rare-earth element. The glass has a lower refractive index than normal glass, giving an image with excellent colour and minimal chromatic aberration. Lanthanum eyepieces have five to eight lenses, depending on the focal length, all of which have blackened edges which create superb contrast and field definition across the field of view. Lanthanum eyepieces of shorter focal length have a built-in Barlow lens ahead of the main lens group, giving good 20mm eye relief for all focal lengths (Figure 3.3). The lanthanum is the modern observer's mid-price deep-sky eyepiece of choice, delivering magnificent views down to f/4.5 across a very flat field free from aberrations.

Wide-Field Eyepieces

Eyepieces described as 'wide-field' contain at least six elements. They give apparent fields of 60° and larger, offering the observer impressive views of the heavens.

The Erfle (Figure 3.4a) was the standard wide-field eyepiece for many years, and with good reason – with its six elements it has an apparent field of 60°–65° and excellent eye relief, all at a modest cost. The downside is that lower-power Erfle eyepieces suffer from considerable outer-field astigmatism, and with the introduction of newer designs from other manufacturers they are no longer the observer's first-choice wide-field eyepieces.

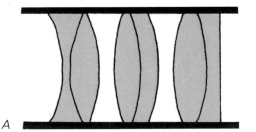
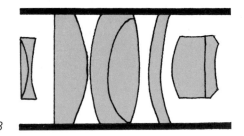

A   B

Fig. 3.4 *(a) The Erfle eyepiece, the standard wide-field eyepiece.*
(b) The Nagler eyepiece, an ultra-wide-field design.

Fig. 3.5 *A standard 31.5mm (1¼in) eyepiece and a 50mm (2in) eyepiece.*

Fig. 3.6 *A standard 31.5mm (1¼in) focuser and a 50mm (2in) focuser.*

If a 60°–65° apparent field is large, then the next generation of super-wide-angle eyepieces are awesome, offering the observer an apparent field of at least 80°–85°. Viewing the heavens through one of these eyepieces is like being there. The first super-wide-field eyepieces were the Naglers (Figure 3.4b). They use a complex arrangement of lenses to achieve superb edge-of-field definition, and work well even with telescopes of short focal length. These big-brand eyepieces offer the deep-sky observer as much for their money as possible, and are highly recommended if your budget allows. You will probably need only one ultra-wide-field eyepiece with an apparent field of 80°–82° to observe the more extended deep-sky objects, and be able to rely on other eyepieces for more detailed views – but bear in mind that wide-field eyepieces are likely to have a 50mm (2in) fitting (see below).

50mm Eyepieces

For many years the standard diameter of an eyepiece was 31.5mm (1¼in), and this is still the standard for the majority of telescopes. However, as the aperture of amateur instruments increases, low-power wide-field views are more in demand, and the 50mm (2in) eyepiece is now a common accessory in the observer's eyepiece case (Figure 3.5).

These eyepieces give a wider field than do smaller-diameter eyepieces: a 31.5mm eyepiece will always display a narrower field than a 50mm. Of course, to use a 50mm eyepiece one needs a focuser of the correct diameter, and most mid-price telescopes now come with a 50mm focuser as standard, complete with a 31.5mm adapter (Figure 3.6).

4 What Else Do I Need?

Acquiring a telescope and some eyepieces is not the end of the story. There are many accessories that make your observing much more enjoyable and productive. Some of the accessories described in this chapter you will probably acquire with your telescope, while others you may need to replace or acquire separately.

Barlow Lens

A Barlow lens is a diverging lens that can be placed between the main optics and the eyepiece of a telescope to increase the effective focal length of the instrument and thus boost

Fig. 4.1 A 6 × 30 and an 8 × 50 finder scope.

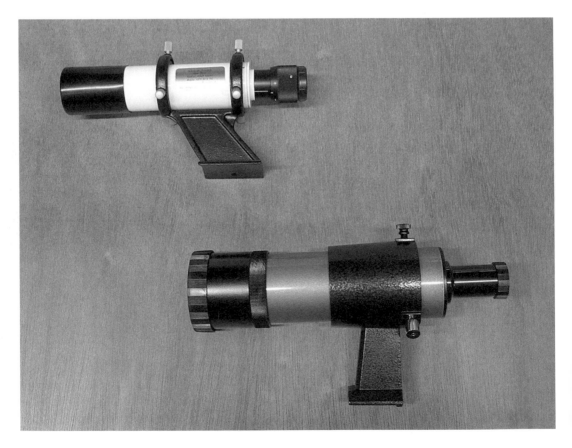

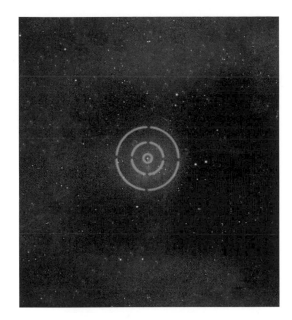

Fig. 4.2 The view through a Telrad, a zero-magnification device used in conjunction with a standard finder.

have a 6 × 30 finder, which is sufficient for finding the brighter deep-sky objects, but a size of 8 × 50 is a more practical for star-hopping (*see* Chapter 7) because the larger aperture resolves fainter stars (Figure 4.1). A different type of finder popular with many observers is the Telrad, a zero-magnification instrument which is a simple pointing device that projects a red bull's-eye onto the sky (Figure 4.2). The Telrad is normally used in conjunction with a standard finder.

Similar in concept to the Telrad is the red-dot finder. This is an easy-to-use, precise zero-magnification finder that works with all astronomical telescopes, and allows even the novice user to locate celestial targets. It superimposes a variable-brightness LED (light-emitting diode) red dot on the sky at infinity that can be accurately and easily collimated with the main scope with just two adjuster knobs. Unlike conventional finders, the red-dot finder works well over a range of viewing distances and can be used with both eyes open.

the magnification. For example, an eyepiece that magnifies by ×50 used with a ×2 Barlow will then magnify by ×100.

If you are planning to use a Barlow lens, take this into account when choosing a range of eyepieces, or you may acquire duplicate magnifications. A 20mm eyepiece used with a Barlow will effectively make the eyepiece a 10mm, so there would be little point in having a separate 10mm eyepiece, and a 12.5mm may be a better choice. If you decide to purchase a Barlow lens, go for the best optical quality you can afford: many manufacturers produce three-element (apochromatic) Barlows, which are vastly superior to their two-lens (achromatic) cousins.

Finder Scopes

All telescopes have a finder – a smaller, low-power instrument attached to the main telescope to help zero in on the area of sky you want to observe. Smaller telescopes usually

Fig. 4.3 A typical dew shield for a refractor.

Dew Shield

A dew shield is a long extension that fits over the front of a telescope to prevent dew from forming on the optics during the course of the night. It also prevents stray light from entering from the telescope tube (Figure 4.3).

Fig. 4.4 An SCT's exposed optical window is
vulnerable to dewing up.

A refractor, by design, is equipped with a
dew shield, but most are not actually long
enough for their intended purpose. A dew
shield should extend forward for at least one
and a half times the telescope's aperture if it is
to provide effective protection. Catadioptric
telescopes such as the SCT have an optical
window at the front of the telescope (Figure
4.4), which is vulnerable to dewing over very
quickly, and for this design commercial heated
dew caps are available.

A Newtonian reflector's main mirror is at
the bottom of the telescope tube, so the tube
itself acts as a very long dew shield. However,
the smaller flat mirror is located very near the
top of the tube and so is susceptible to fogging
over, and also to stray light entering the tube.

Making a dew shield is simple: some thin,
matt-black card rolled to the outside diameter
of the telescope tube and secured to it is all
that is needed. Alternatively, a plastic waste-
paper bin of the right diameter will suffice,
provided the inside is painted matt black to
prevent reflections.

The best solution is to prevent any dew
from forming in the first place by using a
commercial dew-zapper (Figure 4.5). Dew
heaters consist of a low-voltage heating coil
powered by a 12-volt battery. The coil is

Fig. 4.5 A commercial dew-zapper.

Star Diagonal

A star diagonal is a device that fits into the focuser of a refractor and diverts the light path through a right angle to allow a more comfortable observing position. Depending on the diameter of your eyepieces, you will need either a 31.5mm (1¼in) or a 50mm (2in) star diagonal, though most 50mm diagonals have a 31.5mm adapter (Figure 4.6).

Different makes of diagonal differ in quality. Most 31.5mm diagonals use a prism to bend the light path, which can sometimes scatter light, whereas most 50mm diagonals use an aluminized flat mirror. In addition, many 50mm diagonals have enhanced multi-layer dielectric coatings, which give up to 98 per cent reflectivity compared with the 89 per cent reflectivity of standard coatings.

Fig. 4.6 31.5mm (1¼in) and 50mm (2in) star diagonals, shown with a 31.5mm adapter.

Fig. 4.7 Minus-violet filters to counteract chromatic aberration in achromatic refractors.

wrapped around the telescope tube where the objective or corrector lens is; smaller heaters can also be used with finders and eyepieces. The heater provides just enough heat to prevent dew from forming, yet not enough to give rise to tube currents, which may distort the image.

Another popular dew-zapping device is a portable hair dryer, which can either be run from a 12-volt battery, or plugged into the cigarette-lighter socket in your car.

Filters

Filters help the observer to see detail in astronomical objects that may otherwise be unobservable. Filters of interest to the deep-sky are nebula filters and minus-violet filters. Nebula filters transmit only selected emission lines from the spectra of emission nebulae, cutting out wavelengths from street lighting and other artificial sources of light. Minus-violet filters (Figure 4.7) are intended for use with achromatic refractors, which suffer from some

Fig. 4.8 A focal reducer.

chromatic aberration, or false colour, which results in a blue halo surrounding brighter objects. This unfocused blue light reduces contrast and detail in the image, but is rectifiable with a minus-violet filter, which cuts out some of the blue part of the spectrum. (For more about these filters, *see* Chapter 9.)

Coma Corrector

Coma is a common optical aberration found in reflectors of short focal length and causes flaring – star images at the edge of the field elongate into comet-like shapes. This is easily rectified by using a coma corrector, which dramatically reduces coma in f/3.5 to f/6 Newtonian telescopes.

Focal Reducer

A focal reducer threads onto the rear cell of any Schmidt–Cassegrain telescope and reduces the telescope's focal ratio by a factor of 0.63. For example, an f/6.3 focal reducer used with a 200mm (8in) f/10 SCT will reduce its normal 2000mm focal length to 1260mm. With a focal reducer in place, a 25mm eyepiece normally providing a magnification of ×80 will magnify by only 50×. This is useful if a deep-sky object doesn't fit into the field of view at f/10: fitting a focal reducer will shorten the focal length of the instrument, reducing

the magnification of the eyepiece and giving a wider field of view. A focal reducer also helps to correct for any optical aberrations because it reduces field curvature and coma at the edge of the field, ensuring that stars appear as point sources across the entire field of view (Figure 4.8).

Vibration-Suppression Pads

Deep-sky observing often uses high magnifications, so telescope stability is a critical factor. An accidental jolt to a telescope being used with a high power will induce vibrations that will obliterate any detail in the image, totally ruining the view until the vibrations die away. If the telescope is used in windy conditions, it will

Fig. 4.9 A vibration-suppression pad in place.

Fig. 4.10 *The drive system of a driven telescope mounting.*

be impossible to observe anything other than at low powers. Vibration-suppression pads can be used with any tripod-mounted telescope, and when placed between the tripod legs and the ground (Figure 4.9) significantly reduce any vibration caused by an accidental knock to the telescope tube or tripod.

Red LED Torches

When an observer is shielded from any light, chemical reactions in the eye make it more sensitive to dim lighting conditions, a process called dark adaptation (*see* Chapter 1). At the telescope, you should use only a red-light source for illumination, because red light does not destroy dark adaptation. Whatever light source you use does need to provide a sufficient level of brightness – not enough to dazzle the eye, but not so dim as to make writing notes difficult.

Any torch with some sort of red filter attached will serve the purpose, though most tend to be too bright for astronomical use. An alternative is to buy a special observing torch that utilizes one or more LEDs. The battery life

is much longer than with a standard torch, LEDs do not fail as bulbs can, and an LED provides true monochromatic light, fully preserving your night vision.

Clock Drive

A clock drive is a motor attached to an equatorial mounting to enable automatic tracking of the stars (Figure 4.10). When you observe at a high magnification with a telescope on an undriven mount, any object will pass through the telescope's field of view in a few seconds, and you will constantly have to recentre the object in the field. With a driven mount, if the instrument is properly aligned so that the polar axis points to the celestial pole,

Fig. 4.11 *A typical hand controller for a clock-driven mounting.*

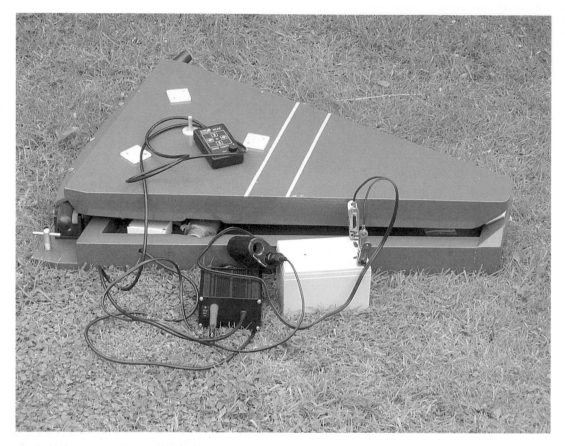

Fig. 4.12 The Poncet platform enables a Dobsonian telescope to track the stars.

the object will stay in the centre of the field of view. Most drive systems also come with a hand-held unit that enables the observer to slew around the sky faster than normal in right ascension and declination (Figure 4.11).

Clock drives are designed for use with an equatorial mounting, so you cannot fit one to a Dobsonian telescope, which is meant primarily for simple low-power visual use, but you can get a Dobsonian to track the stars by using a Poncet platform.

Poncet Platform

The Dobsonian telescope is a Newtonian reflector mounted at its base in a simple altazimuth mount. This design allows Newtonian reflectors of a considerable size to be mounted and still be portable; the downside is that the Dobsonian mounting is purely an altazimuth system and cannot track the stars. The Poncet platform (Figure 4.12) is a special base that the Dobsonian mounting sits on, and once it is set for the observer's latitude and hooked up to a 12-volt battery it will track the stars very accurately for up to an hour before needing to be reset.

5 What Else Can I Observe?

Deep-sky objects fall into three basic categories: clusters, nebulae and galaxies, each of which contains various sub-types. Many deep-sky observers also like to observe double-star systems – stars linked by gravity and orbiting each other, or stars that appear close together on the sky but in reality are separated by vast distances (*see* Chapter 12).

Star Clusters
Open Clusters

Open clusters populate the spiral arms of our galaxy and are collections of gravitationally bound stars. They range in size from a few dozen members barely distinguishable from the background starfield to clusters of several hundred stars a few tens to perhaps a hundred light years in diameter.

Open clusters vary considerably in age. Some, such as the Pleiades (M45) in Taurus (Figure 5.1), are relatively young at 50 million years, young enough still to have traces of the surrounding nebulosity from which the stars were born; NGC 6791 in Lyra, is an example of an old open cluster, with an age of over 600 million years.

A few open clusters, such as the Pleiades, are so close that they are resolvable with the unaided eye. Binoculars begin to resolve individual stars in some of the larger clusters, seen against a fuzzy background of unresolved stars. Telescopes with an aperture of 150mm (6in) or more will fully resolve most clusters if they are viewed at a magnification of around ×25 per inch of aperture (×150 for a 150mm aperture).

Fig. 5.1 A star cluster – the Pleiades (M45).

Globular Clusters

Globular clusters are huge spherical concentrations of stars orbiting high above the plane of the galaxy, some of which are thought to be among the oldest objects in the universe. Globular clusters vary in distance and size. The northern hemisphere's best globular, M13 in Hercules, is 25,000 light years away and around 120 light years across, and contains about 300,000 stars. In comparison, NGC 2419, called the Intergalactic Wanderer, is

Fig. 5.2 A globular cluster – Omega Centauri (NGC 5139).

200,000 light years distant. Observers in the southern hemisphere have the finest example of a globular cluster in the entire sky – Omega Centauri. Lying about 17,000 light years away, it is one of the closest objects of its class to the Earth, also one of the largest with a diameter of at least 150 light years and containing over a million stars (Figure 5.2).

Some globular clusters are visible with the naked eye if you know just where to look; they resemble slightly fuzzy stars. Binoculars will show many of the brighter objects, but only as nebulous disks: they will not reveal individual stars. What you see through a telescope depends on what the aperture is. A 100mm (4in) reflector will just begin to resolve the extreme outer areas in the bigger, brighter globulars such as M13, but more compact objects will not be resolved. Larger apertures will resolve many globular clusters to the core, giving an almost unreal three-dimensional impression.

Nebulae

Nebulae are regions of gas – mostly hydrogen – and dust that populate most of our galaxy. They range in size from a few light years for objects such as planetary nebulae (e.g. M76 in Perseus, with a diameter of 2 light years) to tens or even hundreds of light years (e.g. the Orion Nebula, M42, which is 40 light years across). There are several types of nebula, all having different characteristics and looking very different through the telescope.

Emission Nebulae

An emission (or diffuse) nebula is a huge cloud of highly tenuous gas, chiefly hydrogen, that shines by its own light. Ultraviolet radiation from nearby hot stars causes the gas in the nebula to become ionized and thus shine. There are over four hundred diffuse nebulae in the sky. A few are visible with the unaided eye, many are well seen with binoculars, and under good conditions a telescope will reveal in the brighter examples a wealth of intricate detail. The best known is the Orion Nebula (M42), 1,600 light years distant and visible with the naked eye as a fuzzy star in the sword of Orion (Figure 5.3). At the heart of the nebula is a small group of stars, the brightest of which, Theta Orionis, is one of the stars that make the nebula shine.

Dark Nebulae

Dark nebulae appear as areas of the sky which contain few or no stars. The dust they contain absorbs light, dimming or blotting out anything that is behind the nebula, so these objects are more noticeable if they are situated in front of a bright source such as a Milky Way starfield or bright nebula. Not all dark nebulae are equally dark; they are assigned a value on a scale of

Fig. 5.3 A diffuse nebula – the Orion Nebula (M42).

1–6 to indicate their opacity, 6 being the most opaque. This type of nebula can have any shape, ranging from long elongated areas such as the great rift in the Cygnus region of the Milky Way, or the more compact Coalsack in the southern constellation Crux, to the small but distinctive Horsehead Nebula in Orion (Figure 5.4).

Reflection Nebulae

Reflection nebulae are clouds of dust that reflect light from a nearby star (Figure 5.5). This happens if the nebula contains dust consisting of particles the size of those in cigarette smoke, or if the star is not sufficiently hot enough to ionize the gas and make it emit light. Reflection nebulae are more difficult to observe than other types of nebulae.

Planetary Nebulae

Planetary nebulae were first described by William Herschel in 1782. He named them for their unique appearance though the telescope – these small greenish glows reminded him of the planet Uranus. But the term 'planetary nebula' is misleading as these objects have nothing to do with planets. They consist of a hot star with a temperature of 100,000 degrees kelvin surrounded by a shell of ionized gas resulting from the ejection of the star's outer atmosphere in the latter stages of its life. The star quickly evolves into a white dwarf at the centre of this expanding shell of gas which, over the course of a few tens of thousands of years dissipates into space.

Planetary nebulae as seen through a telescope vary in size and shape. They can appear as nebulous disks; some have an hourglass shape, like the Dumbbell Nebula (M27) in Vulpecula; others have the classic

41

smoke-ring shape, like the Ring Nebula in Lyra (Figure 5.6); some appear simply as a star-like point of light.

If you are trying to find a planetary nebula whose appearance is stellar, you first have to pick it out from the other stars in the field – easy if there are only a few field stars, but a challenge if the object lies in a dense starfield. One method is to use a nebula filter (see Chapter 9) and 'blink' the nebula by holding the filter between eye and eyepiece and moving it back and forth, quickly alternating the filtered and unfiltered views. The nebula will seem to blink on and off, as the filter

enhances the appearance of the nebula while dimming the surrounding stars. Alternatively, you could look out for the give-away bluish colour that often betrays a stellar planetary nebula, but whatever method you choose, you will need a good star atlas such as *Uranometria 2000.0* (*see* Chapter 6).

Supernova Remnants

Supernova remnants are the remains of massive stars that end their life by blowing themselves apart in a supernova, one of the most violent events in the universe. The remains of a star that has undergone such an explosion is scattered through the surrounding region, forming delicate strands of nebulosity that gradually disperse into space. The most

Fig. 5.4 A dark nebula – the Horsehead Nebula in Orion.

Fig. 5.5 (above) A reflection nebula – the Merope Nebula (NGC 1435).
Fig. 5.6 (right) A planetary nebula – the Ring Nebula (M57).

famous supernova remnant is the Crab Nebula (M1) in Taurus, the remains of a supernova witnessed by Chinese astronomers in the year AD1054.

Most objects of this type are too faint for binoculars, but telescopes of sufficient aperture

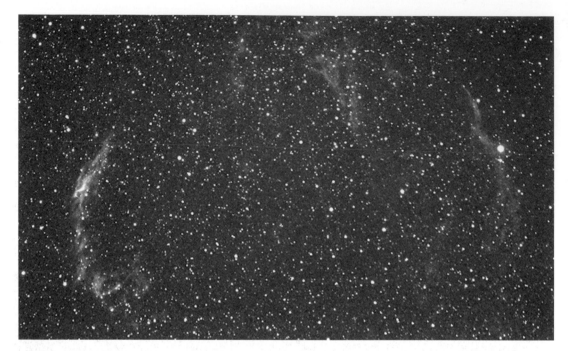

Fig. 5.7 A supernova remnant – the Veil Nebula.

often reveal beautiful, wispy ghost-like structures, especially if a nebula filter is used (*see* Chapter 9). For example, parts of the Veil Nebula in Cygnus, which consists of a huge bubble of material, are visible in small telescopes in a dark sky (Figure 5.7).

Galaxies

Galaxies are the most numerous deep-sky objects. Well over a thousand of these fascinating objects listed in the NGC and IC catalogues (*see* Chapter 6) are brighter than 13th magnitude, within the reach of a modest-sized telescope. But the vast majority of galaxies are no more than misty patches of light in amateur telescopes, and their often low surface brightness makes their internal structure difficult to pick out unless viewed through large apertures.

A galaxy's surface brightness is derived by dividing its magnitude by its area, and is a more reliable indication than its magnitude of its visibility with a given aperture or from a given location. In most deep-sky catalogues the magnitude listed for a galaxy is usually its integrated magnitude. This is the magnitude it would have if all of its light were concentrated at a star-like point. Nearly all galaxies appear in the telescope as extended objects, which is to say that their light is spread out over a certain area of sky. They will therefore have a low surface brightness, and will appear to the observer much fainter than their integrated magnitude might suggest.

For example, the spiral galaxy M33 in Triangulum has a visual (integrated) magnitude of 5.7, on which basis it is a naked-eye target. But this value is predominantly the magnitude of its bright nucleus, which is smaller and more compact than the extended outer regions as seen in a telescope. Viewed through a telescope, M33 will appear much larger as more of its outer regions, which are spread out over a much larger area than the nucleus alone, become visible. In all this galaxy is over one degree across, so these

Fig. 5.8 An elliptical galaxy – M87.

outer regions appear much fainter than the integrated magnitude suggests – as low as magnitude +14 – and if viewed in light-polluted skies these areas may be invisible, only the brighter nucleus being visible.

Generally, the larger a galaxy's dimensions, the lower its surface brightness. This is why very small galaxies only a few arc minutes across are often easily visible in a telescope: their light is not as spread out as it is for larger galaxies, and they have a higher surface brightness. A larger galaxy, though theoretically easier to locate because of its size, has its light is spread out more making it visually much fainter.

Galaxies are not all uniform in appearance: a telescope will show that some are oval, some

Fig. 5.9 (top left)   A spiral galaxy – the Andromeda Galaxy (M31).
Fig. 5.10 (lower left) A barred spiral galaxy.

circular, and some even appear as long thin slivers of light. This is because galaxies are viewed at different orientations, and also they are of four different types: elliptical, normal spiral, barred spiral and irregular.

Elliptical Galaxies

As their name suggests, elliptical galaxies are symmetrical in shape, lacking any sign of structure. They differ from other galaxies in having practically no interstellar gas and dust, and are dominated by old giant stars. The best elliptical galaxy for small apertures is M87 (Figure 5.8), at the heart of the Virgo Cluster of galaxies at a distance of 55 million light years. This is one of the largest galaxies known, with an estimated mass of several trillion Suns, and is several times as massive as an average galaxy. Photographs reveal a huge jet of material shooting out from M87; this jet is several thousand light years in length and is from a supermassive black hole at the galaxy's centre.

Spiral Galaxies

Spiral galaxies are systems whose stars are arranged in a spiral pattern. They come in two varieties: regular spirals and barred spirals.

Fig. 5.11 An irregular galaxy.

Regular spirals have arms that extend from a central nucleus (Figure 5.9). They vary in that some have tightly wound arms, while others have a looser spiral structure. In barred spiral galaxies the arms extend from a central bar straddling the nucleus (Figure 5.10); they show a similar variation in how tightly the arms are wound.

Irregular Galaxies

Irregular galaxies have no symmetry and appear as structureless, often uniform areas of light in a telescope (Figure 5.11).

6 Star Atlases, Catalogues and Software

Atlases

The star atlas is an essential item in the deep-sky observer's arsenal: without one, it would be impossible to find anything of interest – unless, of course, you have a telescope equipped with GO TO technology and can locate objects at the touch of a button (*see* Chapter 7). It is important to have the right kind of atlas for your level of experience. The novice would struggle with an advanced atlas because of the large number of objects plotted: a chart in one of these atlases may only cover an area only slightly larger than the field of view of a typical finder scope. Similarly, the seasoned observer would find an atlas aimed at the beginner frustrating because of the limited number of objects plotted.

Star atlases range from small pocket versions showing only the brightest stars to large hardbound books in several volumes in which a million objects are plotted. The level of detail depends on the atlas's limiting magnitude – the magnitude of the faintest stars it includes. An atlas that plots stars down to 6th magnitude will show over 8,000 stars across the entire sky, and the scale of the individual charts will be sufficient to show only a few hundred deep-sky objects. An atlas with a limiting magnitude of 9.0 will show around 300,000 stars and perhaps 10,000 deep-sky objects. (Star atlases show deep-sky objects much fainter than the limiting magnitude of the stars they plot.) Full details of all the atlases, catalogues and software mentioned in this chapter are to be found in the bibliography at the end of the book.

Atlases for Beginners

For the beginner there are many fine atlases with a limiting magnitude of 5.0 to 6.5 (6.5 being the magnitude of the faintest stars visible to the naked eye under ideal conditions) which are packed with other useful information too.

The *Monthly Sky Guide* by Ian Ridpath and Wil Tirion covers the sky in fifty maps. The book has a chapter for each month of the year and is an easy-to-use handbook for anyone wanting to identify constellations and the main star clusters, nebulae and galaxies. Most of the deep-sky objects discussed are visible to the naked eye, and all can be seen with binoculars or a small telescope.

Terence Dickinson's *NightWatch* is made for use in the field: it is spiral-bound, so that the pages lie flat. The charts each cover a reasonable field of view, and show the most interesting deep-sky objects visible in small- to medium-sized telescopes.

Wil Tirion's *Bright Star Atlas* is a recommended basic star atlas covering the entire sky to 6th magnitude over ten charts. Facing each chart are tables of interesting objects to observe which are shown on that chart.

Slightly more detailed but still in the beginner's category is Wil Tirion's *Cambridge Star Atlas 2000.0*. This is purely an atlas of the night sky and is divided into three sections: monthly sky maps for the northern and southern hemispheres, star charts, and all-sky maps divided into twenty overlapping charts. The charts are in full colour, and the limiting magnitude is 6.5. They show about 900 non-stellar objects, such as clusters and galaxies, which can be seen with binoculars or a small telescope.

The *2000.0* in the title of this and other atlases and catalogues indicates its epoch. The slow change in the orientation of the Earth's axis causes a gradual shift in the positions of stars as measured using celestial coordinates. Accurate positions of objects therefore have to be referred to some standard date – the epoch. '2000.0' stands for noon on 1 January 2000. The previous epoch in wide use was 1950.0. What that means for the observer is that the RA and declination for any celestial object for epoch 1950.0 will be slightly different to the coordinates for epoch 2000.0, but this will be a problem only if one takes a position in 1950.0 coordinates and tries to plot it on an atlas whose coordinates are for the epoch 2000.0. This would have to be taken into account by observers using setting circles to find objects, but star-hopping would not be affected (*see* Chapter 7).

Probably the best-known and most-used beginners star atlas is *Norton's Star Atlas and Reference Handbook*, first published in 1910. The 20th edition, which has the title *Norton's 2000.0,* is a medium-format atlas covering the entire sky in sixteen charts, and plots over 8,800 stars to magnitude 6.5, plus 600 deep-sky objects. It has the added bonus of being a full reference to everything astronomical and has lists of objects that are plotted on the maps. Unfortunately, early copies of *Norton's 2000.0* had printing errors which affected the usability of the charts, and some copies suffer from poor binding. You might be better off seeking a second-hand copy of the 19th edition.

Intermediate Star Atlases

A typical intermediate atlas plots stars to magnitude 8.5 – and so depicts over 36,000 more stars than *Norton's 2000.0*. It has a much larger scale than a beginner's atlas, so the charts don't appear cluttered. In this category there is really only one contender: Wil Tirion's *Sky Atlas 2000.0*. It is large, with a page size of 27 × 39cm (11 × 15in), and the charts double in size when unfolded – which is worth it as each chart depicts a huge 40° × 60° area of sky. The magnitude limit is 8.0, which makes it is very usable with a typical 8 × 50 finder scope, and it plots many thousands of deep-sky objects.

The deluxe version of *Sky Atlas 2000.0* has the sky represented in full colour: the Milky Way is in two shades of blue, and each class of deep-sky object has its own colour. The only concern with this superb atlas is that because of its size it can be a bit of a handful when used in windy conditions. But it is also available in a field edition which consists of smaller individual charts printed on heavy card with white stars on a black background to help preserve night vision at the telescope.

Advanced Star Atlases

If an intermediate star atlas looks detailed, then an advanced atlas looks awesome. There are only two publications which fall into this category: *Uranometria 2000.0* and the *Millennium Star Atlas*.

Uranometria 2000.0 is a hardcover atlas in two volumes. Volume 1 covers from the north celestial pole to declination –6°, and volume 2 from declination +6° to the south celestial pole. This atlas has a larger scale than *Norton's* and *Sky Atlas 2000.0*: here 1° on the sky is equivalent to 1.85cm (about ¾in) on the page. Such a large scale can accommodate fainter stellar magnitudes, enabling *Uranometria 2000.0* to reach magnitude +9.5 and thus 300,000 stars, sufficient to allow the user to star-hop easily with an 8 × 50 finder, plus over 10,000 deep-sky objects – enough to provide the serious observer with years of study.

The *Millennium Star Atlas* is a monumental and expensive (around $250) work, consisting of three hardback volumes covering the entire sky in 1,542 pages. The coverage of each chart is 5.4° × 7.4° on a scale of 4.3cm (1.7in) to 1°. In all the atlas plots over a million stars, twenty-five times more than *Sky Atlas 2000.0* and three times more than *Uranometria 2000.0*. The limiting magnitude is 11.0, faint enough for star-hopping through the actual telescope using a low-power eyepiece.

Deep-Sky Catalogues

All the atlases mentioned above plot deep-sky objects. It is all very well being able to find a particular object on a chart, but if you don't know anything about its visibility you may have a fruitless search. For example, a particular galaxy may appear on a chart but it but be too faint to see in your telescope. If you knew this before looking for the galaxy, you would realize that it was a pointless exercise. So first you need to look up an object in a catalogue. Most star atlases have tables listing the brighter deep-sky targets such as the Messier objects and brighter NGC objects (*see* below) plus lists of double stars and, usually, variable stars.

Essentially a catalogue assigns numbers to deep-sky objects and gives details about them: typically, their size, magnitude, orientation and position, and perhaps a short paragraph on what you might see through different-sized telescopes.

Perhaps the most widely cited listing is the Messier catalogue, containing 110 objects – the so-called Messier objects, mostly the brighter deep-sky objects and perfect for a small telescope – and also a perfect starting point for the novice. Numbers in this catalogue are prefixed with the letter M. Another widely cited catalogue is the NGC (*New General Catalogue*, together with its two *Index Catalogue*, or IC, supplements), containing over 8,000 objects, including all those in the Messier list. (Most objects have entries in different catalogues, for example

M1, the Crab Nebula, also has the designation NGC 1952). The NGC lists objects which are challenging targets for large amateur telescopes, and is a good reference to use in conjunction with a more advanced star atlas such as *Uranometria 2000.0*.

There are many other catalogues, and most are available for download via the internet. Some list only a specific type of object, such as galaxies or double stars, but whatever you decide to observe there will be a catalogue with all the necessary information available.

The best sky catalogues are the ones compiled to accompany a specific star atlas, because they list every deep-sky object plotted on the charts. *The Sky Atlas 2000.0 Companion*, for example, features descriptions and data for all 2,700 star clusters, nebulae and galaxies shown in *Sky Atlas 2000.0*. Its main section lists each star cluster, nebula and galaxy by designation, along with descriptive remarks by well-known observers, and cross-references them to their chart numbers. A second section tabulates the objects and their properties by chart number. Both sections provide coordinates, object type, constellation and apparent magnitude.

Users of *Uranometria 2000.0* need not feel left out because there is *The Deep-sky Field Guide to Uranometria 2000.0*. The DSFG, as it is often called, lists every deep-sky object page by page as it appears in *Uranometria 2000.0*. So if you are observing objects plotted on chart 112 of *Uranometria*, you simply look up chart 112 in the DSFG, where you will find a complete listing of every object plotted on that chart with details such as position, apparent size and surface brightness.

Computer Software

Even the most basic astronomical software will cover more stars and, usually, deep-sky objects than an intermediate star atlas will, and software designed for serious amateur

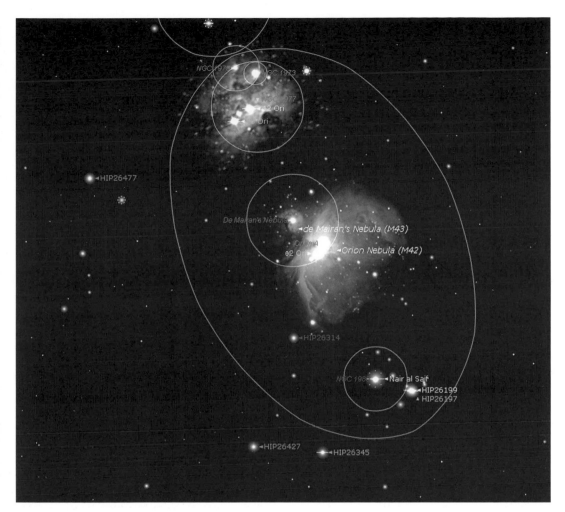

Fig. 6.1 Starry Night Pro. *(© Imaginova Canada Ltd. All rights reserved.)*

astronomers provides information to which a decade ago only professional astronomers would have had access.

Software can be categorized into three areas: sky-simulation programs, star-atlas programs and databases (though most astronomical software has a database facility). The rest of this chapter describes just a small selection of titles in the three categories, and any of them will prove useful to the amateur astronomer.

Sky-Simulation Software

This type of software accurately depicts a realistic night sky simulating such things as horizon profiles, twilight glows, the Milky Way and light pollution. The various titles on the market all do essentially the same thing, the only difference is that graphically they look different.

Starry Night is available in versions from 'enthusiast' to 'professional'. The enthusiast version is more for the novice observer, but still plots 2.5 million stars, more than the *Millennium Star Atlas*, and over 170 deep-sky objects. The professional version of *Starry Night*

(Figure 6.1) plots over 150 million stars and over 100,000 deep-sky objects – enough to keep the serious observer busy for quite a while – and provides access to even more stars and deep-sky objects via the internet. Many deep-sky objects are represented as colour CCD images, while other objects are plotted as outlines at the correct scale and orientation. *Starry Night* also provides full GO TO telescope control (see Chapter 7) via a third-party plug-in.

Similar to *Starry Night* is *The Sky* (Figure 6.2), available in three versions according to budget and skill level: 'student', 'serious astronomer' and 'professional'. The student version plots over a million stars to 12th magnitude, while the serious and professional versions plot over 19 million stars to 15th magnitude. With a wealth of additional stellar databases available, the professional version can display up to a billion stars and over 100,000 deep sky objects. Like *Starry Night, The Sky* uses CCD images to display many of its deep-sky objects, giving a more realistic view of the sky, and GO TO telescope support is available.

A cheaper alternative is *Redshift*. While not as visually appealing as *Starry Night* or *The Sky,* it covers 20 million stars and 70,000 deep-sky objects, and also display objects as CCD images. Although *Redshift* simulates twilight there is no light-pollution facility, and the Milky Way is represented less realistically than in *Starry Night* and *The Sky*.

Star-Atlas Programs

This type of software presents the night sky more as a traditional printed star atlas, such as *Uranometria 2000.0*, would. There are no realistic horizons or light pollution.

With its uncomplicated layout, *SkyMap Pro* (Figure 6.3) is a good example. It shows stars to 15th magnitude and over 200,000 deep-sky

Fig. 6.2 The Sky.

Fig. 6.3 SkyMap Pro.

objects. This software also supports GO TO telescopes equipped with digital setting circles.

Another star-atlas program is *MegaStar*, which plots over 200,000 deep-sky objects from its own database plus others via internet downloads. The software offers full GO TO control.

More unusual is *SkyTools*, which is a complete observer's toolkit. It is primarily a huge database of astronomical catalogues which you can search, using a large number of criteria, to compile personalized observing lists. Once you have made your observing list you can print out customizable finder charts for the naked eye and for finder-scope and telescopic fields of view. *SkyTools* can display either a photorealistic night sky, as sky-simulation software does, or can show a more traditional star-atlas look. The

software can be updated via the internet, where you can download new catalogues or upload your own observing lists for other observers to use. It supports telescopes with GO TO technology. The charts in this book were produced with *SkyTools*.

The above software is commercial, but there is an excellent free title downloadable from the internet called *Cartes du Ciel*, which gives an excellent rendition of the night sky (Figure 6.4). It makes use of the data in sixteen catalogues of stars and nebulae and enables you to prepare different sky maps for a particular observation. The software is very user-definable: you can select, for example, which catalogues to draw upon, which types of deep-sky object to display (based on criteria such as position angle and size), how labels and coordinate grids are displayed, how pictures are superimposed and conditions of visibility.

As well as their value in familiarization with deep-sky objects and the starfields in which they are located, and the preparation of printed charts for observing them, the programs described above can be used while observing at the telescope via a laptop computer. With a suitably equipped telescope, you can use the GO TO facility that many of them possess to drive the telescope to any object in the sky, simply by clicking on an object on the computer screen and choosing 'slew to object'.

A big advantage of using astronomy software while at the telescope is that you have more control over the area of sky you are looking at. It is possible to zoom in on any area of sky. You can start with a wide, naked-eye view to get your bearings, zoom into a smaller area for star-hopping, then finally zoom right in for a simulated view as seen through a high-power eyepiece. This is not possible with a single printed chart, unless you prepare multiple printouts for each deep-sky object you wish to view.

Fig. 6.4 Cartes du Ciel, *an invaluable free program.*

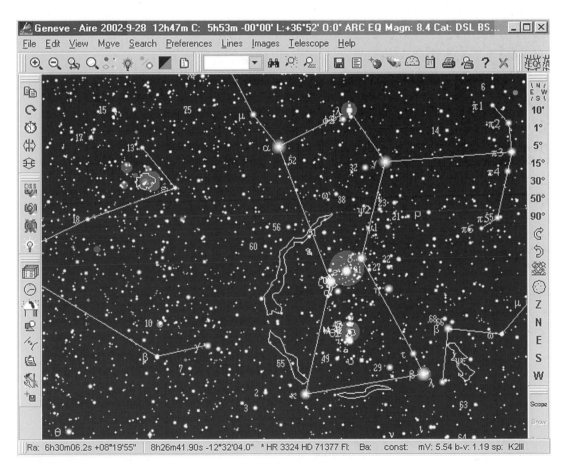

7 Help – I Can't Find Anything!

The sky is clear and there's a new Moon, promising dark skies. You go outside with your new telescope, eager to start observing. You scrutinize a page in the star atlas, choose an interesting object, and pop an eyepiece into the telescope and point it roughly in the direction indicated by the chart. But when you look into the eyepiece you find yourself totally lost, confused by countless stars.

Every amateur will have this experience at some time. Navigating around the night sky is probably the most daunting task the novice deep-sky observer will face, as most objects are very small, just fractions of a degree across – and the night sky is a large place. There are several ways to find your way around the sky, from simple star-hopping to using telescopes with GO TO technology. By far the simplest and least expensive way is to star-hop to your intended target.

Star-Hopping

With star-hopping, you use a star atlas to select a prominent star fairly close to your target, then work out a route to the target that consists of a number of 'hops' from one star to another. Most telescopes have an exceedingly narrow field of view: with a typical low-power eyepiece the telescopic field is only about a degree across, so to point the telescope to the area of sky you are interested in you need to use the finder scope in conjunction with the main telescope. Successful star-hopping requires you to know the field of view of each of your eyepieces and

of your finder, and be able to translate those fields into circles of the correct size on the pages of your star atlas.

Generally, most finders have a field of 5°–7°. You can work out the field by looking at a pair of stars that just fit into the finder's field of view. Northern-hemisphere observers can use the Pointers in the Plough (Big Dipper), the two stars that point to the Pole Star, which are about 5° apart. For southern-hemisphere observers the stars Beta and Nu Octantis are a fraction under 5° apart.

A more accurate method is to select a star near the celestial equator, such as Delta Orionis. Then (with the clock drive switched off, if the telescope has one), time the star's passage as it travels across the finder's field, making sure that the target star passes through the centre of the field. Multiply the result, in minutes and seconds of time, by 15 to find the diameter of the field in minutes and seconds of arc.

Once you have found the finder's field of view, you can draw a circle onto a piece of acetate whose diameter corresponds to the size of the field on the scale of your star atlas, and you can then easily see which stars will be visible in the finder. Star-hopping requires a detailed star atlas that plots stellar magnitudes down to the limit of your finder. Most

Fig. 7.1 The field of view of M27 through (a – above right) a finder and (b – below right) the main telescope. (© Greg Crinklaw, SkyTools 2.)

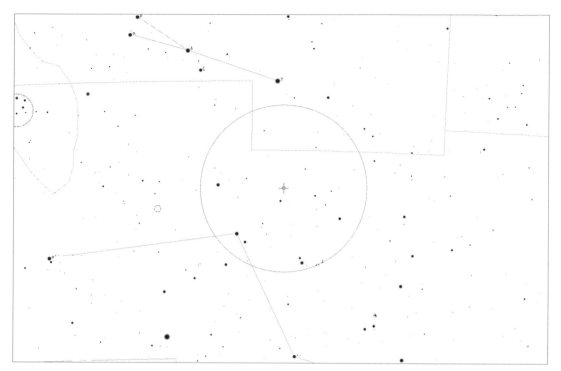

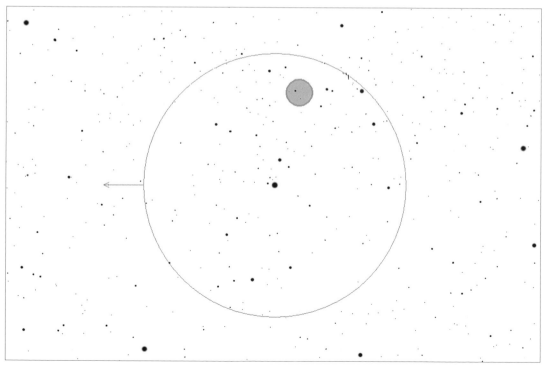

Fig. 7.2 Finding M81 and M82 by the sweep method. (© Greg Crinklaw, SkyTools 2.)

telescopes are equipped with either a 6 × 30 or a 8 × 50 finder, so *Sky Atlas 2000.0* or *Uranometria 2000.0* would be suitable choices.

Finding the Dumbbell Nebula by Star-Hopping

The Dumbbell Nebula (M27), a planetary nebula in Vulpecula, is situated in a rich region of the northern-summer Milky Way that passes through Cygnus. Of magnitude 7.6 and with a diameter of 6.7 arc minutes, it should be visible in good binoculars in good conditions, so is a good choice for this exercise.

The first thing to do when star-hopping is to select a suitable naked-eye star located close to the object of interest. For M27 the easiest star is Beta Cygni, a magnitude 3.4 yellow star marking the head of Cygnus. Then select a second star that will provide an easy hop to the next bright star, and so on, working your way towards the target.

Before you begin, ensure that your finder is aligned with your main instrument by centring a bright star in the finder then looking through the main telescope. The star should be centred in the field; if not, adjust the finder and repeat until the two instruments are parallel with each other.

Start by centring Beta Cygni in the finder (as shown in Figure 7.1a, which is for a finder with a 5° field). Then move the telescope south until the magnitude 4.4 star Alpha Vulpeculae appears in the field. Next, move the telescope east towards 13 Vulpeculae (magnitude 4.6). In the same field, note the star 16 Vulpeculae (magnitude 5.2) about 2.5° north-east, and also 14 Vulpeculae (magnitude 5.7) to the south-east: these stars form a triangle with 13 Vulpeculae, 14 Vulpeculae lying at the apex of the triangle (Figure 7.1a). You are now in the vicinity of M27. Change to a low-power eyepiece and centre the star 14 Vulpeculae in the field – M27 should be visible about 25 arc minutes from 14 Vulpeculae (Figure 7.1b).

Similar methods can be used to track down most deep-sky objects of interest. Look in particular for alignments of stars in an east–west or north–south direction (especially if you are using an equatorially mounted telescope), and pairs, triangles and other groupings of stars that catch the eye.

The Sweep Method

Another simple way to find objects, if you have a telescope on an equatorial mounting, is the sweep method. You centre the field of view on a fairly bright star that has the same declination as your target. Once a suitable guide star is found, the telescope is slowly moved in right ascension away from the star, and will eventually sweep across the object. This method works best if you know the fields of

your eyepieces accurately, because you can then simply measure on your star chart how many low-power eyepiece diameters the target object is away from the guide star.

Finding the Galaxies M81 and M82 by the Sweep Method

To locate these bright galaxies, the magnitude 3.8 star Lambda Draconis (declination +69° 19') is chosen as the guide star. M81 and M82 are at a declination of +69° 04' – not on exactly the same line of declination, but the difference of 15 arc minutes is less than the diameter of a low-power field of view (Figure 7.2). From a star atlas, you can see that Lambda Draconis is 8.5° from M81 and M82. If your low-power eyepiece has a 1° field, you can find the galaxies simply by counting nine full eyepiece-fields across the sky while moving the telescope in right ascension.

Setting Circles

Every equatorially mounted telescope is equipped with setting circles – a set of scales marked on the right ascension (RA) and declination (dec.) axes of the telescope mount.

A telescope's RA setting circle is usually scaled in hours, from 1 to 24, with smaller lines representing 10-minute increments. The

Fig. 7.3 (a) A right ascension setting circle and (b) a declination setting circle.

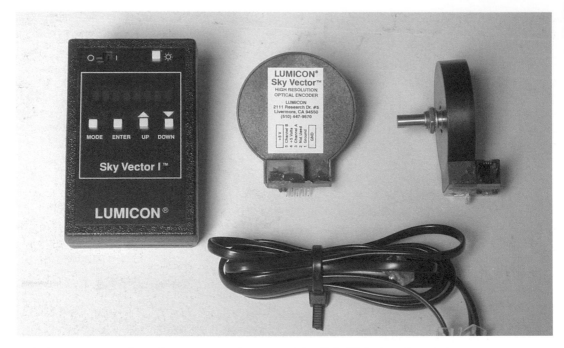

scale has two sets of numbers, one on the top of the scale and another set below, the upper numbers for use in the northern hemisphere and the lower numbers for the southern hemisphere (Figure 7.3a). The declination scale is marked every one or two degrees and runs from +90° to −90°, so one should make sure that you use the correct scale for your side of the equator (Figure 7.3b).

You can use setting circles to find any celestial object simply by looking up its coordinates in a star atlas or catalogue, but you must first ensure that your telescope is polar-aligned and that the RA and declination setting circles are calibrated. To calibrate the circles you must first have a star whose coordinates you know in your field of view. The coordinates of the bright star Vega, for example, are RA 18h 36m, dec. +38° 47'. Centre Vega in the telescope's field of view, unlock the RA and dec. circles on the mount, then rotate both setting circles until they read 18h 36m and +38° 47'. The circles are now ready for use.

Fig. 7.4 Digital setting circles.

Finding the Ring Nebula Using Setting Circles

Once the circles have been set, you can find any object just by moving the telescope while watching the RA and dec. dials on the telescope mount. For example, to find the Ring Nebula (M57) simply look up its coordinates in a deep-sky catalogue; they are 18h 54m, +33° 02'. To zero in on the Ring, move the telescope until the coordinates on your setting circles match the coordinates in the catalogue; once this is done the nebula should be visible in a wide-field eyepiece.

Digital Setting Circles

Digital setting circles can be fitted to almost any telescope, whether it is on an altazimuth or an equatorial mounting. They provide a continuous, real-time digital readout of RA and declination coordinates (Figure 7.4). Digital circles do not move the telescope to its new target automatically, as a GO TO system does – you have to move the telescope manually.

To find an object you first need to choose two stars, preferably on opposite sides of the sky and selectable from the control pad; this is so that the hardware has two reference points and knows where it is pointing. Move the telescope towards the target while watching the coordinates on the control pad as they change while the telescope is moved across the sky. When the readout gives the correct RA and declination of the object, it should be visible in a low-power field of view.

Polar-Aligning Your Telescope

Polar-aligning an equatorially mounted telescope is often regarded as difficult and time-consuming, but it is a simple process and when done will make your observing time much more productive and pleasurable. First adjust the telescope's mounting so that the polar axis points to the celestial pole (the elevation of the pole above your horizon is equal to the latitude of your observing location). This is done by raising or lowering the polar axis of the mounting; all mountings have a latitude scale, so you simply need to set the axis to your latitude. Once a mounting is precisely polar-aligned, the clock drive will track your target: even at high magnifications, the object will stay firmly centred in the field of view.

Go to Telescopes

The ultimate aid for finding objects in the night sky is GO TO technology. A hand-held control pad (Figure 7.5) is connected to encoders which are attached to the right ascension and declination drives of the telescope mounting. GO TO systems allow the observer to select from as many as tens of thousands of objects in the system's database, and the telescope will then whisk off to find that object, centring it in the eyepiece field. Go

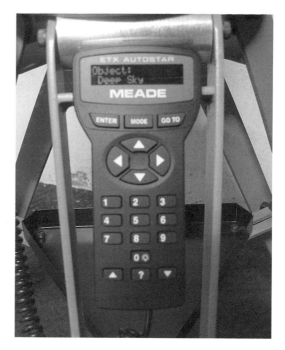

Fig. 7.5 A hand-held controller for a GO TO system.

TO systems contain a virtual map of the sky, complete with accurate coordinates for deep-sky objects (as well as other objects, such as the planets).

Before you can use a GO TO system, you first need an initial alignment of two stars. Accurate polar alignment isn't necessary, but you do need the telescope to be pointing more or less in the direction of the celestial pole. Select an alignment star via the control panel; the telescope will then move to the star's position. Once the drive has stopped, centre the star (if it is not already centred) using the push-button motion controls on the handset, then hit the 'align' button. Now select a second star, preferably on the other side of the sky, and repeat the process. The system will now know where the telescope is pointing, and will be able to slew to any object in its database and accurately track it.

8 Sketching the Deep Sky

Recording what you see through the telescope is an important part of amateur observing. A permanent record of your observing experience will be something you can look back on in years to come or share with other observers to exchange information. There are various ways to record deep-sky objects, such as conventional long-exposure photography and webcam or CCD imaging – all of which will give good results – but for the novice or intermediate amateur it is advisable to begin by sketching what you see, because ultimately this will make you a better observer.

If you are drawing a deep-sky object you will need to do more than just look at it – you will need to observe it carefully, and after a time you will be able to pick out detail you didn't see at first. Of course this does not happen overnight, but the more you observe, the more you will be training your eye.

When you look at a deep-sky object through the telescope, you won't see much at first unless it is something bright, such as an open cluster. If it is a faint or diffuse object you will probably detect only its brightest areas, such as the central nucleus of a galaxy or the brighter, more condensed regions of an emission nebula. Only after patiently staring into the eyepiece will you start to pick out other detail within the object – it takes time for the brain to decipher what the eye is looking at. A galaxy that was merely a fuzzy blob to your inexperienced eye will begin to show a hint of spiral structure. Using averted vision greatly helps the eye to discern detail, and lightly tapping the telescope tube can sometimes help with faint objects because the eye is sensitive to movement. Before starting to observe, make sure that your eyes are fully dark-adapted by staying away from any bright lights for at least half an hour, and then using only red light for illumination at the telescope.

Don't be put off by thinking that sketching is difficult, and requires artistic talent. It doesn't matter if you are not that proficient at putting pencil to paper: you don't need to draw a masterpiece. As long as the dimensions and placement of detail are correct, and the field stars accurately placed, your drawing will serve as a worthwhile record of your observation.

Materials

Compared with other forms of recording, sketching costs next to nothing. All you require are pens, pencils, paper and an eraser. These items are readily available from stationers, but always try to get artist-quality materials, which use pure pigments and give the best results.

The paper you use is important. It is best to use pure white, acid-free sketching paper of a reputable artist's brand; cheaper, non-acid-free paper will gradually become brittle and yellow. A spiral-bound pad is more convenient at the telescope than having lots of loose sheets, but whatever you use try to get paper at least 180gsm in weight, as thicker paper will take more punishment from rubbing and blending techniques.

The correct grade of pencil is also very important, as drawing nebulosity requires

Fig. 8.1 An HB pencil (left) is too hard to represent deep-sky objects, whereas a 3B pencil (right) can be blended nicely.

subtle blending of detail that is barely perceivable to the eye is to be captured. Pencils generally come in two different grades, H (hard) and B (soft), with HB in between. Sketching at the telescope requires the B series, which comes in grades from B to B8, B8 being the softest. Generally, a B pencil and a B3 or a B4 will be fine for astronomical sketching; anything softer just smudges into a black shiny mess. Don't use and HB or H pencil: the H series are for technical drawing, and the graphite is far too hard for sketching deep-sky objects (Figure 8.1).

Even the best artists make mistakes, so you will need an eraser. A soft putty eraser is best as it is pliable, and will rub out without smudging. Finally, for drawing field stars use a black technical pen with a fine nib, 0.5mm in diameter.

Before you start to observe, settle down and make yourself comfortable. Depending what you are going to observe you could be at the telescope for an hour or more. If at first you don't see anything in the eyepiece, don't give up: it takes time for the eye to register the presence of faint objects. Once an object is located, try different magnifications to bring out different levels of detail. For example, at a low power a galaxy may look uniformly bright; only a higher power will bring out interesting detail, because higher magnifications artificially darken the background sky, increasing the contrast of the object.

Sketching Techniques

The technique for sketching a deep-sky object is illustrated here for the Whirlpool Galaxy (M51). Note that deep-sky sketches are made 'in negative', with bright areas of an object drawn dark, and dark areas rendered lighter, for example by using an eraser. If you prefer a more natural image of white objects on a black background, it's

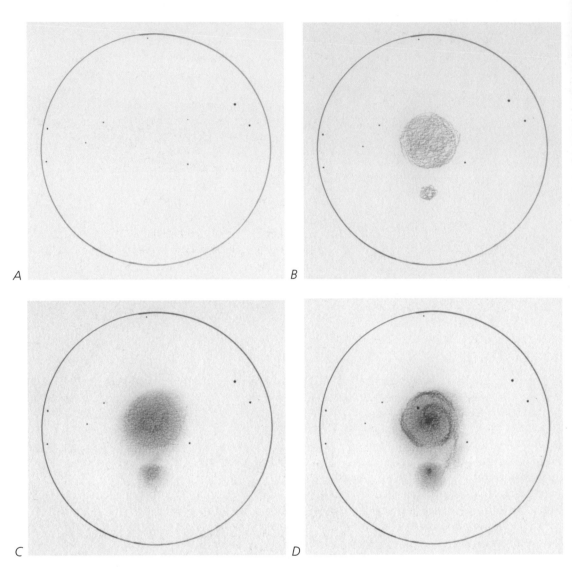

A B C D

easy to scan the finished negative sketch into a computer and convert it to a positive image.

Before starting to sketch an object, draw a circle to represent your telescope's field of view; a good size is about 75mm (3in) in diameter. Then, using the technical pen, position the brighter field stars on the paper, as different sized dots according to their brightness. Take your time to try to position the bright field stars accurately, then draw in the fainter stars in the field (Figure 8.2a).

Fig. 8.2 *Stages in sketching a deep-sky object (the Whirlpool Galaxy, M51): (a) the field stars; (b) the object is roughed in; (c) after blending to give a smooth appearance; (d) the final details added.*

Once the stars are in position, use the side of the pencil point to roughly draw in the shape of the object (Figure 8.2b), then very gently rub the graphite with your finger to blend it and give a smooth appearance. Don't rub the

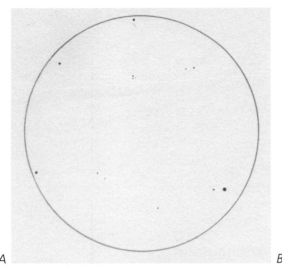

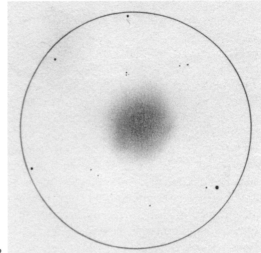

A   B

Fig. 8.3 Stages in sketching a globular cluster
(M13): (a) the field stars; (b) after representing
the concentration of stars; (c) the final details
added.

graphite too much at this stage, otherwise you
will end up with a shiny black mess. It is much
better to apply a few light layers of graphite
than one thick one, slowly building up the
intensity (Figure 8.2c). Continue applying
graphite to the drawing, each time blending it
with your finger, gradually building up the
image until your sketch is a suitable match to
the object as seen in the eyepiece.

Finally, draw in any interesting details you
can see. For example, a brighter nucleus can be
added with the pencil, or the eraser used to put
in any dark dust lanes (Figure 8.2d). This last
stage may require a high magnification. Don't
be in a rush to finish as you may miss important
details. Take your time, and stop only when you
are sure that you have recorded all you can see.

The Whirlpool is a prominent galaxy, of
which there are a number of examples
scattered around the sky. Other galaxies,
though, are not as easy to observe – the vast
majority visible in amateur telescopes appear
only as faint and sometimes barely perceptible

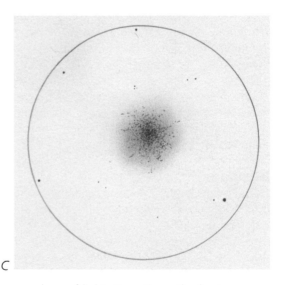

C

smudges of light. Sometimes the best you can
do is just making a visual sighting.

Open Clusters

Open clusters are straightforward to sketch, as
they consist only of stars, with no extended
areas as with galaxies and nebulae. The only
difficulty can be with the complexity of large
clusters, of perhaps several hundred members:
just take your time, and persevere (*see* Figure
11.9, page 92).

Globular Clusters

Globular clusters are quite easy to sketch. There are far too many stars to accurately position them all, so you have to simplify what you see by concentrating on the distribution of stars around the periphery of the cluster. When trying to resolve the individual stars in a globular, always use as much magnification as the conditions will allow. Start by positioning any field stars with the technical pen (Figure 8.3a). Then, using the pencil, draw a circular glow and lightly blend it with your finger to represent the concentration of stars: brighter (darker in the sketch) at the centre, and dimmer (lighter in the sketch) in the outer regions (Figure 8.3b).

The appearance of globular clusters can vary a lot depending on the aperture of the telescope. Small telescopes don't gather enough light to resolve their individual stars, so they appear as fuzzy balls of light. Large apertures will resolve the brighter examples down to the core. Use the technical pen to make the centre of the globular granular, and then position the outer stars, careful noting features such as clumping of stars (Figure 8.3c).

Nebulae

The different types of nebulae can appear very different in the telescope, and there are different approaches to sketching them.

Emission nebulae can be very difficult to draw, especially when viewed in a large telescope or through a nebula filter, as the complexity can be staggering. The best way to sketch this type of object is to use the same method as for galaxies – so be prepared to spend a long time sketching something like M42.

Planetary nebulae are a mixed bag. They are easy to sketch in small apertures as they usually appear only as small puffballs or even just stellar points, but in larger apertures, especially if a nebula filter is used, the detail can be immense. Sketch them as you would a galaxy,

or as an open cluster if they appear stellar (*see* Figure 11.15, page 105).

If a nebula filter (*see* Chapter 9) is used to observe emission or planetary nebulae, note that, depending on what type of filter is being used (UHC, OIII or H-Beta), stars can be dimmed by up to five magnitudes, making field stars difficult to observe unless the filter is removed.

Reflection nebulae are usually faint and don't show much detail in a typical amateur-sized telescope, so use the same method as for sketching a galaxy (*see* Figure 11.17, page 108).

Writing Notes

When a sketch is finished, it is good practice to write a few notes describing the visibility of the object on a preprinted observing form (Figure 8.4). Include information about your equipment, such as the instrument, its aperture, its focal ratio and the eyepiece you used, and also the seeing, transparency, and the date and the time in your time zone.

Seeing is a term to describe the steadiness of the atmosphere, and is rated on a scale of 1–5 called the Antoniadi scale:

I	Perfect, without a quiver.
II	Slight undulations, with moments of calm lasting several seconds.
III	Moderate, with larger air tremors.
IV	Poor, with constant troublesome undulations.
V	Very bad, scarcely allowing a rough sketch to be made.

The seeing determines what kind of observations the observer can undertake: it would be of little use planning a series of double-star observations if the seeing was at the lower end of the scale, because the stars

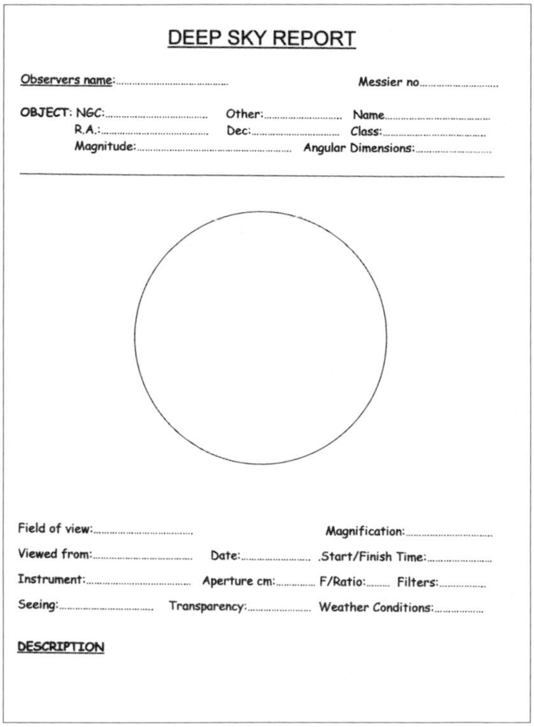

DEEP SKY REPORT

Observers name:................................... Messier no...................

OBJECT: NGC:................... Other:................ Name...................
R.A.:............................ Dec:................. Class:...................
Magnitude:............................ Angular Dimensions:...............

Field of view:........................ Magnification:...................

Viewed from:........................ Date:............... Start/Finish Time:...................

Instrument:........................ Aperture cm:............ F/Ratio:....... Filters:...............

Seeing:................... Transparency:.................. Weather Conditions:...............

DESCRIPTION

Fig. 8.4 *A typical deep-sky observing form.*

would twinkle violently, making any high-magnification observation pointless. When the air is steady, perhaps I or II on the Antoniadi scale, would be a good time to observe close, difficult double stars in which one component is much fainter than the other.

On nights like these the transparency will be good, also allowing excellent views of faint diffuse objects such as galaxies and nebulae. Sky transparency rates the clarity of the atmosphere in terms of the faintest naked-eye star visible at the zenith. Even a little atmospheric haze will affect the visibility of diffuse objects dramatically – on nights like these, nebulosity seems to merge into the background murk.

When describing a deep-sky object, always try to make a size estimate. This is straightforward if you know the true field of view of your eyepieces (*see* Chapter 3). It is also important to state the orientation of the object that you have observed. To do this, you first need to know where west is within your eyepiece field. All astronomical objects rise in the east and set in the west, so simply turn off any drives and note where the object exits the field of view. This point is west; it is then easy to determine the other compass points within the field.

When you write a description of the object you have observed, note what you can actually see. Mention any areas of brightness (condensations) in the object; also mention any other details you see, such as a stellar nucleus in a galaxy or any dark lanes. For an open cluster, estimate the number of stars and note any doubles or any colours in the stars.

9   Light Pollution

Deep-sky objects tend to be faint and diffuse, and require dark skies away from artificial lighting to be seen at their best. Observers who try to track down these objects from a large town or city will find that they are lost in a diffuse glow of orange light from street lights, domestic security lights and other forms of artificial lighting (Figure 9.1). This unwanted glow in the night sky is called light pollution, and it seriously affects the visibility of deep-sky objects. In today's bright night skies, many deep-sky wonders are either totally lost or appear as pale, washed-out patches of light (Figure 9.2).

What Is Light Pollution?

Light pollution has many sources. The spectrum of an urban sky usually shows a continuous glow, spanning the entire visible part of the electromagnetic spectrum. Superimposed on this back-ground glow are bright lines, which come from the airglow and from high- and low-pressure sodium street lighting.

The dark-adapted eye perceives even the darkest night sky as a very dark grey. This is because it is lit by the airglow, which is caused by the emission of sunlight by ions and molecules in the Earth's upper atmosphere. The emission lines of the airglow are at 456, 558, 630 and 636nm.

Low-pressure sodium street lighting emits only very narrow lines in the yellow part of the spectrum, at 589 and 590nm. High-pressure sodium lighting is the most common form of street lighting – and for the astronomer it is the worst kind. These lights emit a wider band of wavelengths, ranging from 498nm in the green part of the spectrum to 569 and 589nm in the yellow-green and 590 to 616nm in the orange.

Types of Lighting

Although some forms of light pollution are completely natural, such as the airglow and moonlight, the vast majority is artificial and is caused by badly designed and inefficient street lighting, lighting from commercial premises such as filling stations, poorly positioned domestic security lights, and even huge searchlights advertising some fancy nightclub.

Tungsten Lighting

The most familiar type of lighting is the standard tungsten light bulb (incandescent lamp), the sort we all use in our homes. At one time this type of lighting was used to light our streets, but it was replaced by mercury-vapour lighting, though there are still places, mostly small villages, where this form of lighting can still be found (Figure 9.3). Tungsten lighting produces a yellowish-white glow that does not significantly intrude into the night sky.

Fluorescent Lamps

Fluorescent gas-discharge lamps work by passing an electric current through a gas or vapour so that a luminous arc is established within a glass tube coated internally with a layer of phosphor. As the current passes through, the tube glows brightly with a very bright white light. This form of lighting is used

Fig. 9.1 (above) A typical light-polluted urban sky.
Fig. 9.2 (right) Deep-sky objects appear washed out in light-polluted skies. Here, M65 (top) and M66 in Leo are barely recognizable as galaxies.

domestically, particularly in kitchens. It seriously degrades night vision, so if you happen to look at a fluorescent light, then be prepared to spend a long time regaining your night vision. This also applies to television screens and CRT computer monitors: the phosphors in the TV and monitor tubes can spoil your dark adaptation. The LCD/TFT screens on modern computers and laptops also have a serious detrimental effect on one's night vision. Some astronomical software has a red-screen mode designed to preserve your night vision; if there

Fig. 9.3 An old-style tungsten street light.

is no such option, then a piece of red acetate taped across the screen can be used to good effect.

Low-Pressure Lighting

The mercury-vapour lamp, in which a high electric current is passed through mercury vapour at low pressure to produce a blue-white light, is still widely used for street lighting in America, but in Europe it was replaced by low-pressure sodium lamps. These work in much the same way, except that the gas used is sodium vapour rather than mercury vapour. Typical low-pressure sodium lighting produces a distinct orange glow with two spectral lines very close together near 590nm, but these can easily be filtered out by using a nebula filter (*see below*). In the UK, high-pressure sodium lights have largely replaced low-pressure sodium lights, but you can still find low-pressure sodium lights illuminating some side-streets.

Fig. 9.4 (a – top) Bad design: a typical low-pressure street light. (b – above) Good design: a full-cut-off high-pressure street light.

High-Pressure Lighting

Unfortunately for the British deep-sky observer, high-pressure sodium lights are not only the most common but also the most intrusive. Their emission lines stretch over the entire visible spectrum, totally swamping the important oxygen lines that most nebulae emit. For the British urban observer, this more than anything else is what makes most if not all deep-sky objects vanish from view in an unnatural pall of light, penetrable only by using special filters.

Good and Bad Lighting

Street-light designs vary considerably depending on what type of lamp they use. One type of badly designed low-pressure sodium street light consists of a 360° glass surround (Figure 9.4a), which has the effect of sending much light up into the sky instead of down on the ground, where it is needed. This is a major contribution to light pollution.

High-pressure lighting, though brighter than low-pressure lighting, is actually more convenient for the observer because the lighting fixtures are much better designed. The lamp is located high in the housing, and the beam of light is constrained to shine downwards by reflectors mounted inside the housing. This type of lighting is called full cut-off (Figure 9.4b).

Street lighting is not the only source of the light pollution that the urban observer has to contend with. Probably the most intrusive is the 500-watt tungsten-halogen security light (Figure 9.5), often mounted high on a wall and aimed horizontally to illuminate as wide an area

Fig. 9.5 A badly positioned home-security light.

Fig. 9.6 Ornamental street lighting like this may look attractive, but there is no shielding to protect the night sky against light pollution.

as possible. High-wattage lights are also used to floodlight commercial and public buildings, making them visible for huge distances. Another source is the type of ornamental lighting found in parks, supermarket car parks and outside shopping centres. These lights often consist of a white opaque globe which allows light to spread out in all directions; because they have absolutely no cut-off they are highly inefficient at illuminating the ground, so several are often grouped together (Figure 9.6).

Filtering the Bad Light
There are various ways to beat light pollution. If you have a portable telescope you can travel to a dark country location where the urban glow is less intrusive. If you are affected by a nearby

light you can make some sort of light shield. And by using a good-quality nebula filter (Figure 9.7) you can restore washed-out deep-sky objects to their former glory.

Nebula filters are made by finely grinding and polishing a piece of glass to a perfectly flat optical figure to ensure that no image deterioration occurs, even at high magnifications. Dielectric coatings are deposited on both sides of the glass; these let through the desirable emission lines while blocking all other unwanted spectral lines.

Nebula filters block out any unwanted light, whether it be natural airglow or artificial street lighting. They do this by transmitting only a few specific wavelengths in a small portion of the green part of the spectrum (which is also the peak transmission of the dark-adapted eye, around 510nm) while blocking out the unwanted blue and yellow light on either side. The main visible emission from nebulae is at the wavelength of doubly ionized oxygen (OIII), near 500nm, and hydrogen-beta, which is a faint, very narrow emission line at 486nm.

Nebula filters don't make the object being observed any brighter, they simply make the background darker, enhancing the contrast between sky and object – which unfortunately has a detrimental effect on any field stars. If you use a line filter such as the H-Beta or OIII, you will find that stars are dimmed by up to five magnitudes. Also, if you use a line filter with a smaller telescope, the field of view can become so dark that it can be difficult to see anything.

Types of Nebula Filter
Nebula filters fall into three categories: broadband filters, narrowband filters and line filters. Broadband filters have the widest bandpass, making them suitable for most deep-sky objects, but the wider the bandpass, the less contrast gain is achievable. A narrowband filter is used to observe emission and planetary nebulae; the bandpass is narrow enough to reject most light pollution and is

now regarded as a filter for general use. Line filters are specialist filters designed for observing certain types of object.

Most manufacturers of astronomical equipment pro-duce nebula filters, which are moderately priced. The filters described below are available from Lumicon, but filters of a similar type from other manufacturers will be the same or very close to Lumicon's specifications.

Lumicon Deep-Sky Filter

This is a broad-band light-pollution-reduction (LPR) filter, designed to eliminate low-pressure sodium-vapour and mercury-vapour lighting, plus the emission lines of the airglow. It has a wide, 90nm bandpass (441–535nm). Broadband filters of this type thus give less contrast gain than do narrowband and line

Fig. 9.7 *A set of good-quality nebula filters.*

filters. From my own limited experience of deep-sky type filters, most objects are not significantly enhanced, and the sky background is only marginally darkened. What I find the deep-sky filter best suited for is the viewing of large, face-on spiral galaxies such as M101 in Ursa Major, M33 in Triangulum and M51 in Canes Venatici. In moderately light-polluted skies, using this filter may make all the difference between seeing a galaxy and not detecting it at all.

Lumicon Ultrahigh-Contrast Filter

The narrowband filter is by far the most widely used by amateurs today. It has the capability to cut through quite severe light pollution, pulling out tendrils of nebulosity from the urban glow. The Ultrahigh-Contrast (UHC) filter is a general-purpose narrowband filter. It has a 24nm bandpass at 484–506nm, which encompasses the two OIII nebula lines at 496 and 501nm,

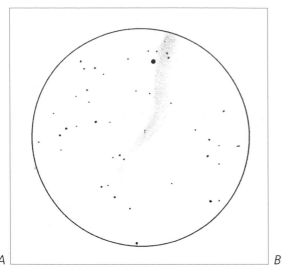
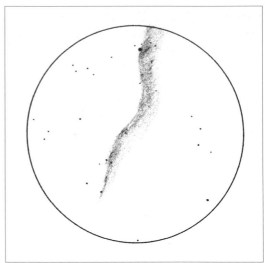

A   B

Fig. 9.8 *The Veil Nebula (NGC 6960) as seen through a 220mm (8¾in) Newtonian at ×65 (a) without a filter and (b) with an OIII filter.*

plus the hydrogen-beta nebula line at 486nm. Since most light pollution is filtered out, the contrast of emission and planetary nebulae is greatly enhanced. Narrowband filters are designed primarily for observing emission nebulae and planetary nebulae. They are not suitable for observing galaxies, star clusters or reflection nebulae as these objects are essentially shining by starlight (reflection nebulae shine by reflected starlight), and narrowband filters kill starlight.

Lumicon H-Beta Filter

This is a line filter with a bandpass of 9nm centred on the hydrogen-beta nebula line at 486nm. Since it lets only hydrogen-beta light through, the H-Beta filter is of little use for observing planetary nebulae. What it is used for is to observe the very faintest emission nebulae – the types that emit a distinct reddish colour on photographs – and other objects such as the Horsehead Nebula (a dark nebula) and the California Nebula (a

reflection nebula). In fact, for some objects the H-Beta filter may be the *only* filter to permit visual observation in badly light-polluted skies.

Lumicon OIII Filter

This is a specialized filter intended for the observation of planetary nebulae. It has an 11nm bandpass which isolates just the two lines of doubly ionized oxygen, at 496 and 501nm, which greatly enhances the contrast of any object that emits light at these wavelengths. This enables you to see planetary nebulae that are not visible with any other filter. The OIII filter also works well with emission nebulae that emit most of their light in the 496–501nm region.

Using an OIII filter is like doubling the size of your telescope: objects stand out in sharp contrast against a very dark sky background, and even in a modest aperture objects such as the Veil Nebula in Cygnus can show so much detail that it can become difficult to make an accurate drawing.

In an unfiltered view the Veil Nebula appears as a very faint, diffuse strip of nebulosity lying in a rich starfield (Figure 9.8a). Its shape can just be discerned, and it is of uniform brightness. Through an OIII filter the Veil is seen as a bright band of nebulosity with well-defined sides; the western side is much brighter (Figure 9.8b). The field stars in the filtered view are much fainter.

Without a filter, the Owl Nebula (M97) is well defined although with slightly diffuse edges in a 300mm (12in) aperture, and the central star is prominent; no other detail can be seen (Figure 9.9a). Viewed through an OIII filter, it becomes a bright ball of nebulosity, and the owl's 'eyes' are easily seen (Figure 9.9b). Again, notice how the field stars are not as prominent as in the unfiltered view, and the central star is difficult to spot with the filter in place.

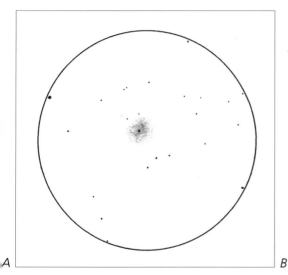 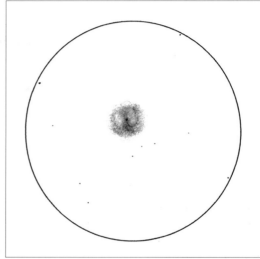

A B

Fig. 9.9 *The Owl Nebula (M97) as seen through a 300mm (12in) Newtonian at ×84 (a) without a filter and (b) with an OIII filter.*

10 Telescope Care and Maintenance

Although telescopes are precision instruments, they are also surprisingly robust and require minimal adjustment or maintenance. Of course, as an amateur observer you would not handle your own telescope carelessly, but there will be times when, however much care you take, the images you receive do not seem as good as they once were.

If objects appear of low contrast or faint, then perhaps the optics are dirty: dirty optics scatter light, reducing the contrast of the image. This can also happen with a Newtonian reflector if the aluminium coatings on the mirrors have tarnished. A Newtonian's optics are exposed to the elements, and are susceptible to dew and dust, which over a period of time slowly eat away their aluminium coatings. As the reflectivity of the coatings diminishes, the image becomes fainter. When this happens the optics have to be removed from the telescope and sent away to be re-aluminized professionally. If the image is blurred or distorted, then the optics are probably out of alignment and need realigning – a procedure known as collimation.

Cleaning the Main Optics
Refractors
Take the utmost care when handling your telescope's optical components, especially if you need to take them out of their housings. A refractor's objective lens should if possible be cleaned without removing it from its lens cell. If the objective does have to be taken apart, you will need to make a note of the orientation of its component lenses with respect to one another. In most refractors these lenses will have a mark on their edge, so when you reassemble the objective you just need to realign the marks. If there are no marks, then before you separate the lenses mark their edges with a soft pencil or black marker pen. In most objectives the lenses are air-spaced by three small wedges of silver paper, and it is vital that you put these back in place when the objective is reassembled.

Cleaning a lens requires cleaning solution (available from any photographic store) and some hospital-quality cotton wool, containing no artificial fibres. Before you touch the lens, remove any dust or grit with a soft blower brush (obtainable from a camera store), which will minimize any chance of scratching the surface of the lens. When you actually clean the lens, use only a minimum of cleaning solution – too much, and you may end up with streaks on the surface – and with the lightest touch. After each stroke discard the cotton wool and take a fresh piece, again to minimize the chance of scratching the lens.

To clean the optical window of a Schmidt–Cassegrain telescope, use the same method.

Reflectors
Cleaning a Newtonian's optics is more time-consuming, but is still a straightforward process. First, remove the primary mirror and the flat from the tube. Unscrew the primary mirror from its cell and examine its condition. It

is likely that there will be bits stuck to the surface; if so, *do not* wipe them off, because that could easily scratch the mirror. Astronomical mirrors are coated with a very thin layer of aluminium. These coatings are easily damaged by careless handling – even the sweat on your fingers will etch itself into the surface of the coatings – and every bit of damage will reduce a mirror's optical performance. Instead, use a soft blower brush or a can of compressed air to remove any loose pieces of dirt or grit.

Once the mirror is free of any loose dirt, take a plastic dish larger than the mirror and clean it thoroughly. Remember, it takes only one bit of grit to severely damage a mirror. Half-fill the dish with hand-hot tap water and add a small amount of liquid detergent. Place the mirror in the soapy water and leave for ten minutes; this soaking period will soften any other stubborn bits of dirt.

After soaking, take a ball of pure cotton wool, raise the mirror to a slight angle, while keeping it submerged in the water, and very gently glide the cotton wool back and forth across the mirror surface in a single direction. After one pass, take a fresh piece of cotton wool and repeat the procedure along a parallel track, working back and forth until the whole of the mirror's surface has been cleaned. Rinse the mirror with cold water and examine its surface; if any dirt remains, soak the mirror in the soapy water for a little longer then repeat the process.

Once the mirror is completely clean, do a final rinse with distilled water. Do not use tap water for the final rinse as tap water contains too many impurities, which could react with the aluminium coating. The final rinse with distilled water will remove any soapy residue from the mirror and wash away any impurities from the tap water. Finally, leave the mirror to drain at an angle. If there are any beads of water left on the mirror's surface after it has dried, dab them with a piece of blotting paper.

Finally, replace the mirror in its cell and reinstall it in telescope tube.

The same method can then be used to clean the secondary mirror.

Eyepieces

Because eyepieces are handled more than any other piece of astronomical equipment, they tend to get a lot dirtier. You always seem to get fingerprints on the eye lens, or traces of grease from your eyelashes as you press your eye against it, which tends to blur the image. All you need to clean an eyepiece is some good-quality lens cleaning solution and pure cotton wool. First, use a blower brush to remove any dirt and grit from the lens, then apply a drop of cleaning solution to the cotton wool – not to the lens itself, as the solution may seep into the lens housing – and very gently wipe it over the lens.

Collimating the Optics

If you know that the optics are clean, and the image is still poor, then probably the optics are out of collimation. An uncollimated telescope will produce blurry, distorted images.

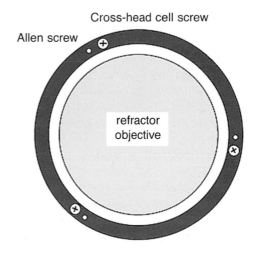

Fig. 10.1 A refractor's adjustable lens cell with adjusting screws.

TELESCOPE CARE AND MAINTENANCE

Refractors

A refractor seldom needs collimating as it will have been collimated properly by the manufacturer, and if it is taken care of it should hold that collimation for its lifetime. However, if you suspect that a refractor is out of adjustment, then it is straightforward to put right. In some refractors the objective is in a sealed cell, so you cannot adjust it yourself; if you suspect that the objective is misaligned you will need to return it to the manufacturer. Other refractors have an adjustable objective cell in which three cross-head screws spaced 120° apart hold the cell together (Figure 10.1). These screws need to be loosened for adjustment to take place. There are also three small Allen screws, one next to each cross-head screw. When screwed in (or out), these small screws will push (or pull) against the main lens cell, tilting it slightly for fine adjustment.

Reflectors

The optics of a Newtonian reflector are much more likely to become misaligned than a refractor's optics because it has two mirrors that are adjustable. If only one of these mirrors is misaligned, it throws the whole optical configuration out of collimation.

The Newtonian's primary mirror is secured to a cell at the bottom of the tube. The adjustment screws are at the back of the cell. Smaller mirrors have three screws (Figure 10.2), and larger mirrors could have as many as eighteen, but the function is the same: to physically tilt the mirror in its cell and change the direction of the light path. The secondary mirror (the diagonal) is held in place by a four-vaned spider, adjustable with three or four screws that tilt the mirror in various directions.

To find whether the telescope needs collimating, in daylight rack the focuser out as far as it will go and, without any eyepiece inserted, look it, paying attention to the positions of the reflections of the mirrors. In a

Fig. 10.2 A Newtonian reflector's adjustable mirror cell with three adjusting screws.

correctly collimated telescope, the reflection of the main mirror is central in the outline of the diagonal (Figure 10.3a). If one or more mirrors are out of alignment you will see something like Figure 10.3b, indicating that the instrument needs collimating.

First, centre the secondary mirror (diagonal) so that it shows the reflection of the entire primary mirror. Do this by loosening the screw and rotating the diagonal in its holder until the reflection of the main mirror is fully visible, then tighten the screws. If the whole of primary mirror's reflection still cannot be seen, you need to turn the adjusting screws on the back of the diagonal's holder to tilt the diagonal until the image of the mirror is central.

Now study the image of the primary mirror. The black outline of the diagonal will be seen, as well as the reflection of the spider that holds the diagonal in place. If the diagonal is offset against the reflection of the primary mirror, you need to turn the adjusting screws on the back of the primary mirror cell to tilt the mirror; depending on how large the mirror is, there could be up to eighteen of these screws to play

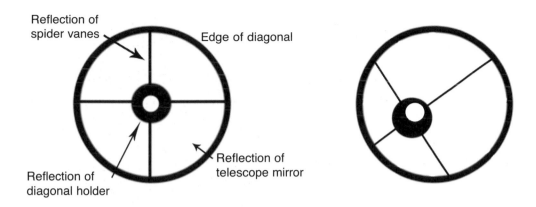

Reflection of spider vanes

Edge of diagonal

Reflection of telescope mirror

Reflection of diagonal holder

Fig. 10.3 (a) A correctly collimated and (b) an uncollimated Newtonian reflector.

with. Adjust these screws until the outline of the flat appears central in the reflection of the primary mirror.

Once the optics are more or less aligned, you can fine-tune by using an actual star. Select a 2nd- or 3rd-magnitude star and centre it in a high-power eyepiece (×200 or higher). Focus the star, then turn the focuser so the image becomes slightly defocused, the star expanding into a small disk. If this disk is non-circular, adjust the primary mirror slightly and

Fig. 10.4 The three adjusting screws on a Schmidt–Cassegrain telescope can be seen at the centre in this view.

then re-examine the star and continue until everything is circular. It is helpful to have someone to assist in this operation: one person adjusts the mirror cell, while the other examines the star image.

Schmidt–Cassegrains

The Schmidt–Cassegrain telescope is similar to the reflector in having a main mirror and a secondary mirror, but the SCT's primary mirror is permanently set by the manufacturer, so the only way to collimate the instrument is by adjusting the secondary mirror (Figure 10.4). To check the collimation, take the telescope outside and centre a 2nd-magnitude star in a high-power eyepiece. Then defocus the star so that you see a bright circle of light (the defocused star) with a circular dark area (the shadow of the flat) inside it. The shadow of the flat should be central in the bright circle. If it is not, turn one of the screws on the secondary while looking into the eyepiece and note where the shadow moves, continue until the disk of the defocused star and the shadow of the flat are central with respect to each other. Once the optics are aligned, fine-tune as described above for a reflector.

11 Touring the Deep Sky

Many thousands of deep-sky objects are visible through even a small astronomical telescope if observed from a dark site, while the number of objects within the grasp larger instruments becomes almost uncountable. The question is, which of these objects are worth looking at? This chapter presents the best and the brightest deep-sky objects, listed by constellation, suitable for binoculars upwards. A few are naked-eye objects, while others will be challenging targets if observed from under a light-polluted sky. Some, although visible in small apertures, will reveal intricate detail if observed through larger instruments.

All the objects described in this chapter can be found in any intermediate star atlas such as *Sky Atlas 2000.0*, and most are listed in *Norton's*, so they should not be too difficult to find. In addition, their coordinates, size and magnitude are given in a table at the end of this book.

Andromeda

Andromeda lies away from the plane of our galaxy, so its deep-sky objects are relatively unaffected by the dust and gas that tends to dim objects near the galactic plane. It is best known for the Andromeda Galaxy, the largest member of the Local Group – a collection of over 30 galaxies within a few million light years of one another, of which the Milky Way is also a member.

M31, the Andromeda Galaxy, is the nearest spiral to the Milky Way, lying at a distance of 2.2 million light years, and is visible with the naked eye as a fuzzy 4th-magnitude star. Standard binoculars such as 10 × 50s show M31 only as a featureless glow (Figure 11.1), and a small telescope shows no more than a featureless, elongated haze with a brighter centre, only slightly better than the binocular view.

In apertures of 150mm (6in) upwards and in good seeing conditions, it *may* be possible to detect a dark lane between the spiral arms, plus a bright emission nebula (NGC 206) in the SW of the galaxy, more visible through a nebula filter. As with all diffuse types of deep-sky object, take your time – the more patient you are, the more you will see. The Andromeda Galaxy has several satellite galaxies, gravitationally bound to it as are the Magellanic Clouds to our own Milky Way.

M32 is the most obvious companion to M31 and is visible in a 60mm (2¼in) refractor as a tiny fuzzy star; larger apertures reveal a circular ball of light, with a much brighter stellar nucleus.

M110 is another satellite galaxy to M31, much more difficult to see than M32 because it is much larger at 19 × 11 arc minutes, giving it a low surface brightness of 13.6 mag/arcmin2. A 150mm (6in) aperture will show a featureless patch of light elongated SE–NW.

NGC 752 is a large, magnitude 6.6 open cluster 49 arc minutes across, so an eyepiece with a

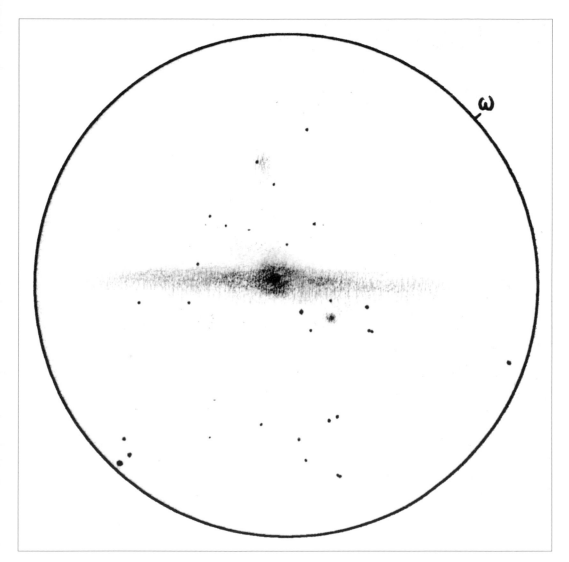

Fig. 11.1 M31 as seen through 16 × 80 binoculars.

field of view of 1° will give the best view. A small telescope will show around 60 stars and a pair of bright yellow stars to the SSW, the brighter of which is 56 Andromedae; there is also a nice orange giant at the centre of the cluster.

NGC 891 is an extremely tough edge-on spiral.

Its integrated magnitude is 10.8, but because of its large size, 14 × 3 arc minutes, it has a surface brightness of 13.8 mag/arcmin2, making it a difficult object in light-polluted skies. A 150mm (6in) aperture will show a thin sliver of light, with perhaps the equatorial dust lane glimpsed in excellent seeing conditions, but a 200mm (8in) aperture is the minimum needed to view the dust lane properly. A magnitude 11.4 star lies at the NW edge of the galaxy (Figure 11.2).

NGC 7662 is an easy planetary nebula for a small telescope. At 17 arc seconds across and shining at magnitude 9.8, it is visible in a 60mm (2¼in) telescope as a non-stellar object. A 150mm (6in) aperture shows a circular blue-green nebula with diffuse edges and a faint magnitude 13.2 star just off its eastern edge (Figure 11.3). The central star, also of magnitude 13.2, is difficult to view because of the high surface brightness of the nebula itself.

Aquarius

Aquarius contains a variety of deep-sky objects: open clusters, globular clusters, planetary nebulae and galaxies, though the galaxies are rather faint.

M2 is a bright, magnitude 6.6 globular cluster, 16 arc minutes across and easily visible in binoculars. A 150mm (6in) telescope shows a grainy ball of light with a 10th-magnitude star

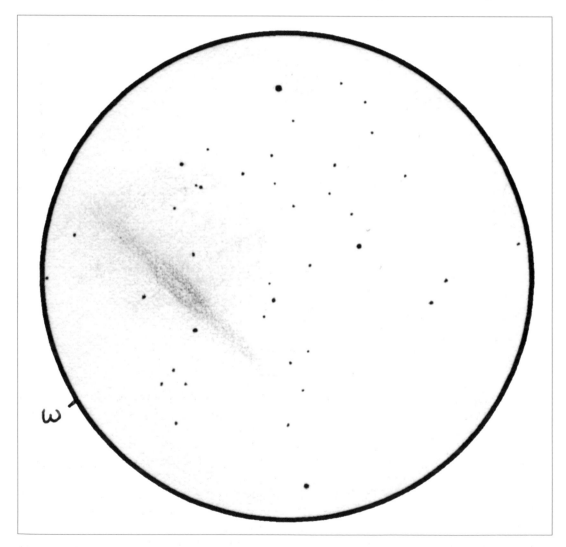

Fig. 11.2 NGC 891, 220mm (8½in) reflector, ×65.

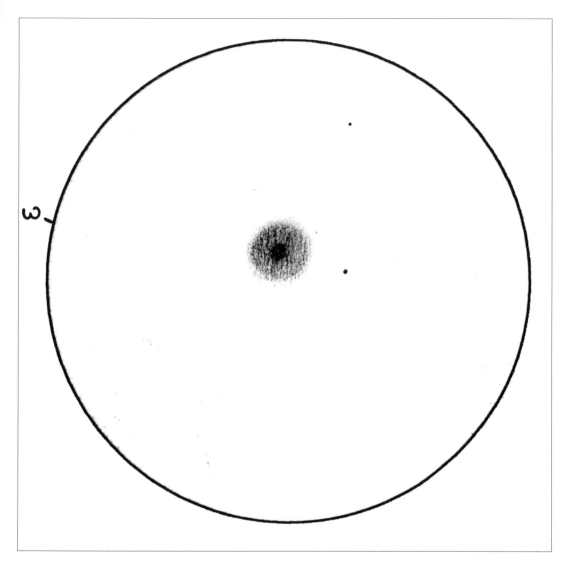

Fig. 11.3 NGC 7662, 220mm (8½in) reflector, ×208–288, UHC filter.

4.5 arc minutes NE of the centre. A 250mm (10in) aperture at high magnification fully resolves this cluster.

M72 is a much more distant globular, about 60 thousand light years away, shining at magnitude 9.2 and 7 arc minutes across. A 150mm (6in) aperture shows a granular ball of light at a magnification of ×100, while a 200mm (8in) aperture reveals the cluster as a mass of stardust.

M73, just under 1.5° west of M72, is simply four 10th- and 11th- magnitude stars in a Y shape. This may not be as visually appealing as most other Messier objects, but is an easy target for the beginner. Messier included it in his list of objects as he thought he could detect some nebulosity around the stars, though

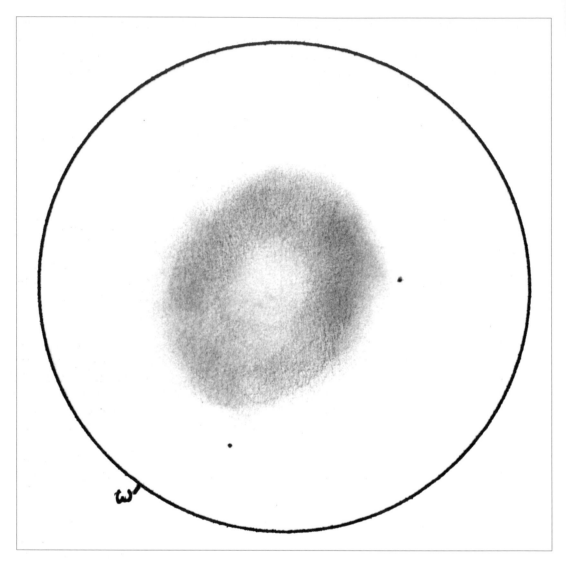

Fig. 11.4 *NGC 7293, 300mm (12in) reflector, ×65, OIII filter.*

modern telescopic observations show no trace of nebulosity. The fours stars are easily visible in any aperture, and in a 150mm (6in) telescope the westernmost of them is seen to have a noticeably orange hue.

NGC 7009 is a magnitude 8.3 planetary nebula 29 arc seconds across. It is called the Saturn Nebula as through an aperture of 150mm (6in) it resembles the planet Saturn, though a high power is needed to see this effect. A 250mm (10in) telescope reveals the extensions on either side of the planetary to be as long as the main oval body itself, and also shows a 14th-magnitude star 45 arc seconds to the NE.

NGC 7293, the Helix Nebula, is a magnificent planetary nebula 1° east of the magnitude 5.2 star Upsilon Aquarii. The Helix is magnitude

7.6, and with a size of 16 arc minutes is visible in binoculars from a dark site; a 150mm (6in) aperture shows a ghostly disk of light about 15 arc minutes across with several fainter stars embedded in the glow. The central star, shining dimly at magnitude 13.4, is visible in a 300mm (12in) aperture (Figure 11.4).

Aquila

Aquila lies in the plane of the Milky Way and therefore contains rich starfields. Most of its deep-sky objects are planetary nebulae, but all are quite small and rather faint, requiring detailed finder charts to pick them out from the background stars.

NGC 6709 is a magnitude 7.4 star cluster 13 arc minutes in diameter, visible as a hazy spot in binoculars. A 150mm (6in) aperture reveals around 45 stars, with a magnitude 9.1 red star a little SSW of the cluster's centre. Larger apertures reveal up to 90 stars down to 13th magnitude.

Ara

Ara is a far-southern constellation south of Scorpius and is never visible for mid- to high-northern observers.

NGC 6204 is a magnitude 8.4 open cluster, 5 arc minutes across. A 60mm (2¼in) refractor will show ten stars, while a 250mm (10in) aperture reveals around 20 stars spread across 4 arc minutes. Also note, 8 arc minutes to the SE, Hogg 22 – a small, 1.5 arc minute cluster consisting of a short chain of stars.

Aries

This dull constellation contains only three moderately bright stars, but as it lies away from the plane of the Milky Way it is rich in galaxies, most of them faint and for large apertures and a dedicated observing eye.

NGC 772 is a magnitude 11.2 galaxy with a low surface brightness of 13.9 mag/arcmin2,

requiring a dark sky for a good view. A 150mm (6in) telescope will show a small patch of light with a brighter core, while a 250mm (10in) aperture reveals that the galaxy is elongated SE–NW.

Auriga

The galactic equator passes through the southern area of this constellation, so many fine clusters and nebulae are visible with telescopes of all sizes.

M36 is a magnitude 6.5 open cluster visible in binoculars as a misty patch of light. A small telescope shows around 20 stars against a hazy background; the total expands to 60 in a 150mm (6in) aperture, with several close doubles visible at the eastern side of the cluster.

M37 is a splendid cluster of magnitude 6.2 visible in binoculars as a misty area; this cluster is quite condensed, so small telescopes reveal only a smattering of stars against a nebulous backdrop. A 150mm (6in) aperture resolves the cluster into a few hundred pinpoints with a prominent magnitude 9.2 red star at the cluster's centre; this star is an easy double, with a magnitude 10.3 companion 20 arc seconds away (Figure 11.5).

M38, another open cluster, is also an easy binocular target at magnitude 6.8 and 21 arc minutes across. Small telescopes reveal about 30 stars, with a 9th-magnitude red star at the centre.

IC 2149 is a small but bright planetary nebula; it is magnitude 11.2 and 8 arc seconds across, so needs a high power to distinguish it from the stellar background. At ×200 magnification in a 150mm (6in) aperture it is visible as a minute disk, detectable for what it is only by its blue-green colour, a common trait of planetary nebulae. The central star, of magnitude 11.6,

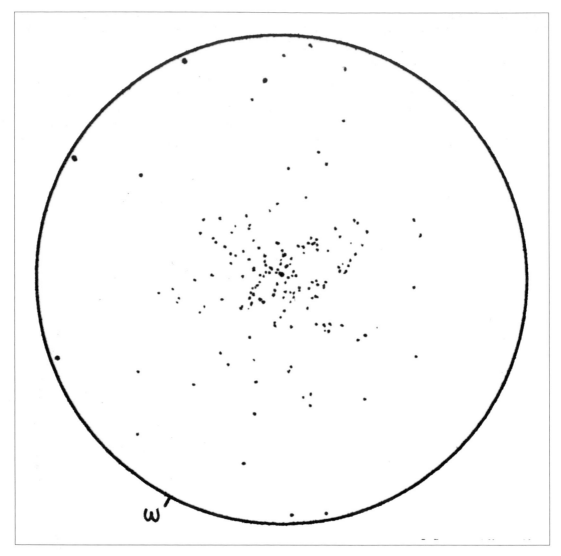

Fig. 11.5 *M37, 220mm (8½in) reflector, ×65.*

is barely visible in larger-aperture instruments because of the nebula's bright centre.

Bötes

A prominent constellation containing the bright orange giant Arcturus, the brightest star in the northern sky, Bötes is populated by faint galaxies.

NGC 5466 is a loose globular cluster of magnitude 9.2 and 9 arc minutes across lying 52,000 thousand light years away. It has a low surface brightness, so is easy to bypass in small telescopes, which reveal only a very weak patch of light. A 200mm (8in) aperture pulls a few individual stars out of its glow, while larger instruments resolve more stars against a granular background.

Camelopardalis

This is a visually barren constellation in the far northern sky, its brightest star being only of 4th magnitude. Galaxies populate this region, but where the constellation's western edge dips into the Milky Way there are some open clusters and planetary nebulae.

NGC 1502 is a bright, magnitude 4.1 open cluster 7 arc minutes across visible in binoculars or a small telescope. The cluster has many double stars.

NGC 2403 is a splendid spiral galaxy of magnitude 8.9, but as its size is rather large at 23 × 12 arc minutes it has a rather low surface brightness, 13.9 mag/arcmin2, so a dark sky is essential when trying to view it. Even a small

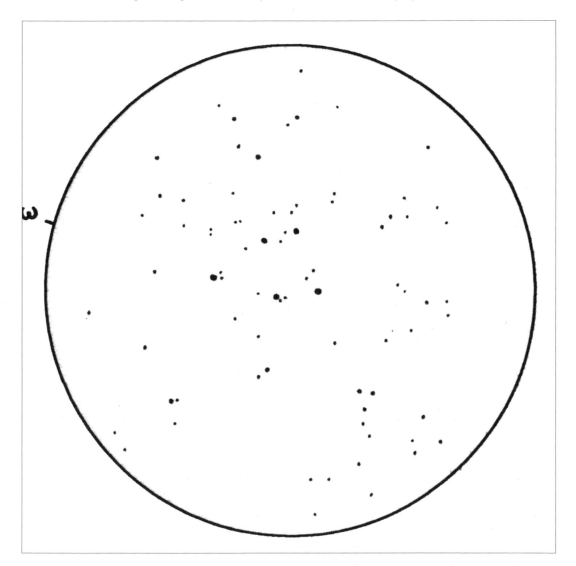

Fig. 11.6 M44, 220mm (8½in) reflector, ×41.

telescope will show a blur 15 arc minutes long, elongated SE–NW; larger apertures reveal stars superimposed on the nebulous glow of the galaxy, plus a hint of spiral structure.

Cancer

This quite faint constellation between Leo and Gemini has a high galaxy content, though none are very impressive in smaller instruments. However, Cancer does possess two marvellous open clusters.

M44, known as Praesepe or the Beehive, is a large open cluster of magnitude 3.9, visible as a naked-eye glow which in good seeing conditions seems to sparkle with individual stars just on the verge of naked-eye resolution. The cluster is 95 arc minutes across – over three full Moon diameters – so the best overall view will be in a pair of binoculars. A small telescope will show about 55 stars, while a 150mm (6in) aperture reveals 75 stars down to magnitude 12. Owners of large-aperture instruments can look for several faint galaxies which lie among the stars of the cluster (Figure 11.6).

M67 is an ancient open cluster in the southern reaches of the constellation; its magnitude is 7.3 and it is 29 arc minutes across. Small telescopes show a partially resolved cluster of about 20 stars against seen a background haze. A 150mm (6in) telescope reveals about 50 stars, while larger apertures show a very rich and compact grouping of stars, among them several old, red stars.

Canes Venatici

This constellation is rich in galaxies, some of which are the finest examples of their kind.

M3, a nice magnitude 6.3 globular cluster 18 arc minutes across, is often regarded as the finest globular in the northern sky. A small telescope shows a granular halo of light; in a

150mm (6in) telescope at high power it is well resolved, and larger apertures fully resolve the cluster.

M51, the Whirlpool Galaxy, is a magnificent face-on spiral galaxy. It has an integrated magnitude of 8.9, but because of its large angular size, 11 × 8 arc minutes, it appears visually much fainter, with a surface brightness of 13 mag/arcmin2. Even so, the Whirlpool is an impressive object, visible in a small telescope as a small circular patch of light with a brighter centre. A 150mm (6in) aperture in good seeing conditions will show a large bright halo with a much brighter middle and a stellar nucleus, possibly with a hint of spiral structure.

200mm (8in) telescopes will show a brighter version of the 150mm view, but give a better chance of detecting any spiral structure. Large amateur telescopes, especially in the 300mm (12in) class, reveal a large bright mottled halo of light, very bright at the centre, like an unresolved globular cluster, and a pinpoint stellar nucleus; the spiral structure is very evident in this aperture. Modest apertures reveal a magnitude 13.5 star well within the halo of the galaxy; also note NGC 5195, a magnitude 9.6 peculiar galaxy, which has a faint bridge of material connecting it to M51 (Figure 11.7).

M63, the Sunflower Galaxy, is an easy object for a small telescope: shining at magnitude 9.3, this 13 × 8 arc minute galaxy appears elongated E–W, and lies just south of a magnitude 8.5 star. In a 250mm (10in) telescope the southern side of the galaxy appears to be flattened.

M94 is a bright, magnitude 8.8 spiral galaxy with very tightly wound arms; visually it looks like an elliptical galaxy. In a 150mm (6in) aperture M94 is a large halo of light with a blazing core, possibly elongated; a 250mm

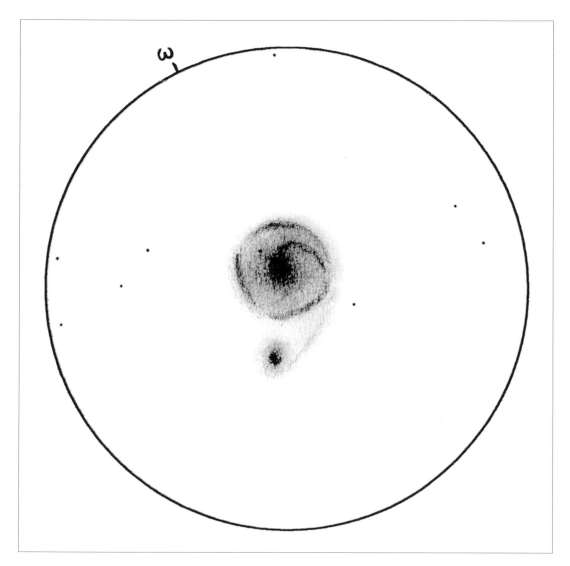

Fig. 11.7 *M51, 220mm (8½in) reflector, ×130.*

(10in) instrument reveals the core to be around 45 arc seconds across with a faint stellar nucleus.

M106 is one of the sky's brighter galaxies, shining at magnitude 9.1. This 17 × 6 arc minute spiral is easily visible in small telescopes as an elongated streak of light; a 150mm (6in) aperture shows it as a longer streak of light

with a much brighter centre and a well-defined core. Telescopes of 300mm (12in) aperture and higher will in *good conditions* reveal some spiral structure (Figure 11.8).

NGC 4449 is a magnitude 10.1 irregular galaxy visible in small telescopes as a fairly bright, smooth-looking glow. A 250mm (10in) telescope will show an arm curving NW, and in good conditions averted vision will reveal a hint of a spiral arm.

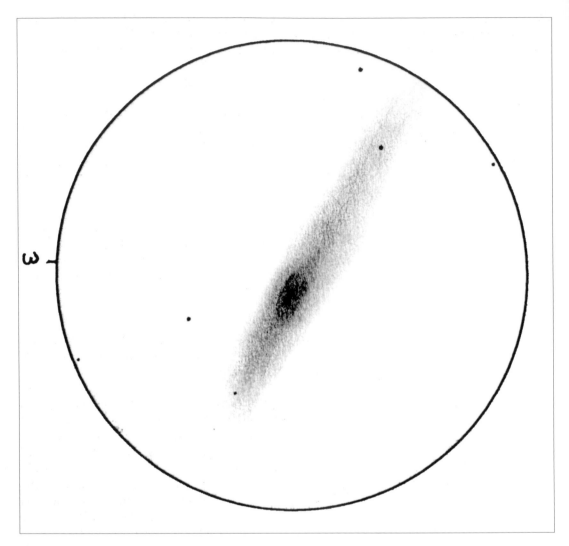

Fig. 11.8 *M106, 300mm (12in) reflector, ×134.*

Canis Major

A constellation just south of the celestial equator, dominated by Sirius, the brightest star in the entire sky, Canis Major lies very close to the Milky Way and is populated by many open clusters and nebulae.

M41, the only Messier object in the constellation, is a magnitude 5.0 open cluster, 38 arc minutes across and visible with the unaided eye. A small telescope shows about 30 loosely arranged stars, including 12 Canis Majoris, in a low-power field. A 150mm (6in) aperture reveals about 50 cluster members, including a few red stars.

NGC 2362 is a magnitude 3.8 open cluster surrounding the bright magnitude 4.5 star Tau Canis Majoris, whose light interferes with the fainter magnitude 9–10 members, but a 150mm (6in) aperture will show around 20–25 stars.

Capricornus

This dull constellation lies in a barren part of the sky and is pretty devoid of anything interesting.

M30 is a magnitude 6.9 globular cluster, 12 arc minutes across, lying less than half a degree from the magnitude 5.24 star 41 Capricorni, which is a fine double with a magnitude 11.5 companion 5.5 arc seconds away. A 150mm (6in) partially resolves the cluster, while a 250mm (10in) resolves it well at high power.

Cassiopeia

This stunning constellation sits in the Milky Way and is packed with open clusters and nebulosity suitable for all apertures.

M52 is a magnitude 8.2 open cluster, 12 arc minutes across, visible in small telescopes as a nebulous patch. A 150mm (6in) aperture shows about 30 members down to 10th magnitude, while larger apertures will show at least 150 stars in the cluster.

M103 is a magnitude 6.9 open cluster, 6 arc minutes across. A small telescope shows the cluster framed in a triangular grouping of magnitude 7.2, 8.2 and 10.5 stars. A 150mm (6in) shows around 20 stars, while a 200mm (8in) reveals around 50 cluster members.

NGC 129 is a sparse cluster visible in small telescopes as a few stars scattered across a half-degree field, while a 150mm (6in) telescope reveals a fairly uniform distribution of stars with a triangle of magnitude 8.9, 8.9 and 9.0 stars to the SE.

NGC 281 is a faint nebulosity 4 arc minutes across nicknamed the Pacman Nebula for its resemblance to the character from the 1980s computer game. It is best viewed by using a nebula filter with a low-power eyepiece. A small telescope shows a low-surface-brightness glow surrounding a small open cluster, while a 200mm (8in) aperture shows NGC 281 as a slightly irregular area with a distinctive bite out of one side, giving it its 'Pacman' appearance.

NGC 457 is commonly called the ET Cluster or the Owl Cluster because of its shape (I think the Owl Cluster is a more suitable name!). This magnitude 5.1 open cluster is 13 arc minutes across and contains the star Phi Cassiopeiae (magnitude 4.9). It is impressive: even in a 60mm (2¼in) telescope about 25 stars are visible; a 150mm (6in) aperture will show about 50 stars, including a few red stars dotted here and there, the most prominent of which is the variable V466 Cassiopeiae (Figure 11.9).

Centaurus

Centaurus is a large southern constellation which unfortunately is unobservable from mid-northern latitudes.

NGC 5128, also known as Centaurus A, is a magnitude 7.7 galaxy 29 × 21 arc minutes in diameter and easily visible in binoculars as a small fuzzy glow just under 5° south of NGC 5139. A small telescope will show a bright circular haze bisected by a dark bar, and through a 150mm (6in) aperture a large, very bright oval area is seen with several superimposed stars, and the dark lane is obvious. Larger telescopes show two bright lobes completely divided by a wide, dark lane.

NGC 5139, also known as Omega Centauri, is the sky's best globular cluster. At magnitude 3.9 and a huge 55 arc minutes across, it is an impressive sight even with the unaided eye. A small telescope reveals a very large, very condensed oval glow which appears granular at low power. In a 150mm (6in) aperture the cluster is well resolved, almost filling a low-power field.

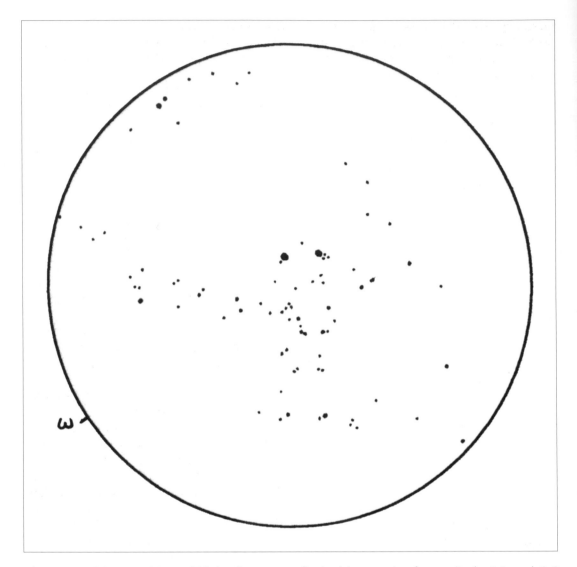

Fig. 11.9 *NGC 457, 220mm (8½in) reflector, ×104.*

Cepheus

A fine northern constellation partially embedded within the Milky Way, Cepheus is populated by many fine open clusters and nebulae.

NGC 40 is a magnitude 10.7 planetary nebula, 48 arc seconds across; small telescopes show it only as a very small disk, even at a high power,

flanked by a pair of magnitude 9.1 and 9.6 stars. Larger telescopes reveal a conspicuous magnitude 11.5 central star. The star flanking the nebula to its NE is a double resolvable in a large telescope, and consists of a red primary (magnitude 9.1) with a 10th-magnitude companion just 0.4 arc seconds away.

NGC 188, an ancient open cluster of magnitude 9.3 and 13 arc minutes across, is a difficult object in a small telescope, visible only as an unresolved glow. The cluster is still faint in a 150mm (6in) aperture, but a smattering of stars is resolved across its face; it takes a 300mm (12in) telescope to resolve the cluster members.

Cetus

Cetus is a large constellation containing a good number of galaxies for the amateur observer.

M77 is a spiral galaxy over 80 million light years distant and shining at magnitude 9.5. This 7 × 6 arc minute object is visible in a small telescope as a bright but small patch of light. A 150mm (6in) aperture reveals a bright, almost circular object with a very bright core and a 9th-magnitude red star just to the NE.

NGC 246, a magnitude 8.0 planetary nebula 4 arc minutes across, is visible in a small telescope as a faint, featureless disk of light. A 150mm (6in) aperture reveals three 11th-magnitude stars superimposed on the disk of the nebula, one which is probably the central star of the planetary.

NGC 247 is an edge-on spiral of magnitude 9.7 and 21 arc minutes across. Its very low surface brightness of 14.1 mag/arcmin2 makes it a difficult object for a small telescope, but even in a small aperture it can be seen as an elongated faint streak of light elongated N–S with a magnitude 9.5 star at its southern tip. Not much more is seen in a 150mm (6in) aperture – it is just brighter and better defined, but a magnitude 10.9 star is visible just off its western side.

Columba

A mid-southern constellation, Columba contains only a single object worthy of amateur instruments.

NGC 1851 is a magnitude 7.1 globular cluster visible as a small granular area in a 150mm (6in) aperture.

Coma Berenices

This is a dull constellation to the unaided eye, but is a treasure trove for the galaxy-hunter: the huge Virgo Cluster of galaxies – also known as the Coma–Virgo Cluster – spills over into Coma Berenices (*see under* Virgo).

M53, a magnitude 7.7 globular cluster less than a degree NW of the magnitude 4.3 star Alpha Comae Berenices, is visible in a 60mm (2¼in) refractor as a fuzzy halo of light with a magnitude 8.9 star near its centre. A 150mm (6in) aperture reveals a partially resolved granular glow, while anything larger on a steady night resolves M53 into a myriad faint stellar points. Owners of larger telescope can look for NGC 5053, which lies a mere 6 arc minutes WNW of M53; this faint, magnitude 9.9 globular needs at least a 250mm (10in) aperture to be seen at all well, and even then it is little more than a dim glow with about 20–25 faint stars sparsely scattered across its face.

M64, the Black Eye Galaxy – so called because of its distinctive dark dust lane – is a magnitude 9.3 spiral galaxy measuring 11 × 5 arc minutes. It is easily visible in a small telescope as a bright oval glow with a brighter nucleus, but to see the 'black eye' a 150mm (6in) aperture will need to be used, with a high magnification. This feature is on the NNE side of the nucleus, and in a 250mm (10in) telescope it is an easy target – a dark patch resembling a kidney bean.

M85 is one of the brightest galaxies in the Virgo Cluster and is easily visible in a small telescope as a magnitude 9.2 glow lying very close to a magnitude 10.6 star. Unusually, larger apertures do not seem to enhance the view, they only make it larger and brighter: it looks like a large, unresolved globular cluster through a 300mm (12in) telescope.

M88 is a giant spiral galaxy with an integrated

magnitude of 10.3, but its low surface brightness of 13.0 mag/arcmin2 makes it a more difficult object than its magnitude suggests. However, even a small telescope will show this galaxy as distinctly elongated, with a faint magnitude 10.6 star visible to the south, while a 150mm (6in) aperture reveals that the star is actually double, while M88 itself is bright but with no detail evident. Larger telescopes show two magnitude 13.8 and 14.7 stars within the halo of the galaxy.

M91 is a magnitude 10.9 barred spiral, but like M88 its surface brightness is low (13.5 mag/arcmin2), making it a harder object than its integrated magnitude suggests. A 150mm (6in) telescope will show the galaxy to be slightly elongated NE–SW, with a smooth texture.

M98 is a magnitude 10.9 spiral, measuring 10 × 3 arc minutes. It is a disappointing sight in a small telescope, visible only as a weak glow, but a 150mm (6in) aperture fares much better, revealing a bright core and a stellar nucleus. A magnitude 13.0 star is visible to the NE of the centre, plus a magnitude 12.8 star superimposed on the galaxy itself.

M99 is a magnitude 10.4 face-on spiral, 5 arc minutes across and visible in a small telescope as an elongated glow extending E–W. A 150mm (6in) aperture shows a brighter halo of light with a brighter mid-region, and the outer area appears granular. Telescopes of 250mm (10in) and larger show an intensely bright core about an arc minute across, and a magnitude 13.2 star on the SW of the halo. Keen-sighted observers with large telescopes can look for a magnitude 14.4 star on the eastern periphery.

M100 is a bright galaxy in small telescopes, visible as a uniform patch of light elongated ESE–WNW; a 150mm (6in) aperture reveals a smooth glow with a small bright core, as do larger instruments – but with the addition of a few faint field stars.

NGC 4559 is a moderately bright, magnitude 10.5 spiral galaxy covering 12 × 4 arc minutes and visible in a small telescope as a bright, elongated area. A 150mm (6in) aperture reveals two stars of magnitudes 11.9 and 12.3 superimposed on the glow of the galaxy.

NGC 4565 is a magnificent edge-on spiral galaxy of magnitude 10.6. It is visible in a small telescope as a dim sliver of light elongated SE–NW. A 150mm (6in) aperture shows it as a long needle of light, slightly mottled in good conditions; the mid-region thickens slightly at the central bulge, and in good seeing conditions there is a hint of the equatorial dust lane, and a magnitude 13 star is visible 1.5 arc minutes NE of the centre. Larger telescopes show an elongated galaxy with a prominent dust lane and central bulge.

Corvus

This small constellation just south of Virgo contains a few interesting deep-sky objects for amateur telescopes.

NGC 4038/39 is a pair of colliding galaxies called the Antennae from their appearance on photographs. Their combined magnitude is 10.9, and they span 6 × 4 arc minutes. A 150mm (6in) aperture will show one bright area about 2.5 arc minutes across, with a magnitude 8.7 star located 6 arc minutes to the NW and a magnitude 11.6 star at about the same distance to the SW.

NGC 4361, a magnitude 10.3 planetary nebula 1 arc minute across, is visible in a 150mm (6in) aperture as a diffuse patch of light; the central star, of magnitude 13, is visible at a high power. Larger telescopes show a bright inner region surrounded by a fainter, more diffuse outer glow extending N–S.

Crux

A far-southern constellation, Crux has a number of open clusters and dark nebulae within its boundaries, two of which deserve mention.

NGC 4755, also called the Jewel Box, is one of the prettiest open clusters in the entire sky; it is of magnitude 5.2. Any telescope will show several dozen stars between 5th and 12th magnitude, and also some of the subtlest colours in any cluster, including the magnitude 5.9 reddish-orange star Kappa Crucis and a very red magnitude 7.2 star close to the cluster's centre.

The Coalsack is a huge dark nebula located at the SE of the constellation not far from Alpha

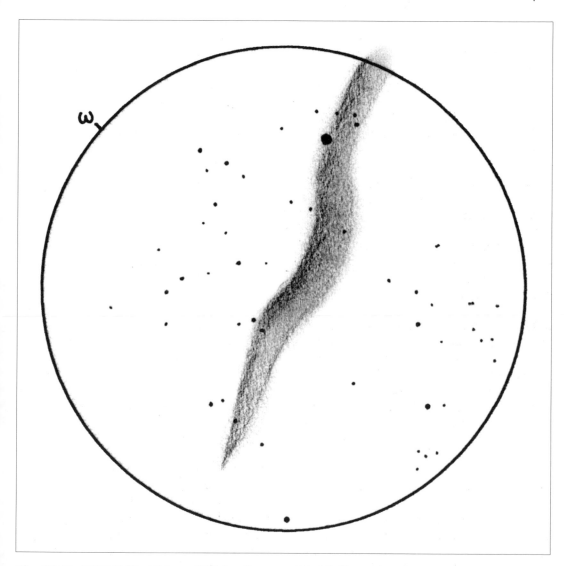

Fig. 11.10 NGC 6960, 220mm (8½in) reflector, ×65, OIII filter.

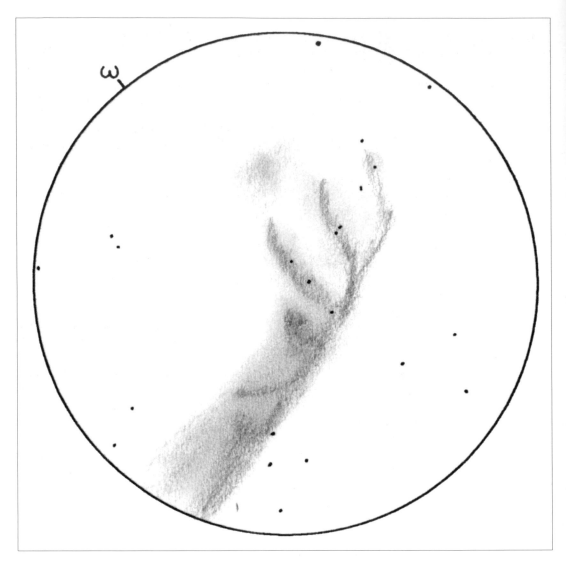

Fig. 11.11 NGC 6995, 220mm (8½in) reflector, ×65, OIII filter.

Crucis. Dark nebulae are rated on a scale of 1 to 6, where 1 indicates barely darker than the surroundings and 6 means the most opaque and easiest to see. The Coalsack has a rating of 6, which with its large size of around 7 × 4° makes it very obvious to the unaided eye. Binoculars, with their wide field and low power; give the most pleasing view.

Cygnus

This magnificent northern constellation lies on the Milky Way and is densely populated with deep-sky objects.

M29 is a small open cluster of bright stars close to the star Gamma Cygni. At magnitude 7.5 it is an easy target for a small telescope, consisting of around 12 stars; a 150mm (6in) aperture reveals about 15 main stars, seven being in a distinctive box shape.

M39 is a large, 31 arc minute open cluster several degrees north of Alpha Cygni (Deneb). Around 20 stars in a distinctive triangular shape are visible in a small telescope, while a 150mm (6in) aperture reveals about 50 stars from 7th to 11th magnitude.

NGC 6826, known as the Blinking Planetary, is a curious little magnitude 9.8 nebula 25 arc seconds across and only half a degree from the star 16 Cygni. A small telescope will show it as a tiny spot of light, while a 150mm (6in) aperture reveals a bright circular nebulosity, with the central magnitude 10.7 star faintly visible at a high power; larger apertures show that the outer edge is more diffuse. This object is called the Blinking Planetary because if you view it directly you tend not to see the nebula itself, only its central star, but look away slightly, using averted vision, and the nebula flashes into view.

NGC 6910 is a bright, magnitude 7.3 open cluster visible in the same field as the star Gamma Cygni in a low-power view. A small telescope shows a dozen stars, with magnitudes 7.0 and 7.4 stars at either end.

NGC 6946 is a very faint, face-on spiral galaxy with an integrated of magnitude 9.7, but a low surface brightness, 13.8 mag/arcmin2 – so wait for a dark, moonless sky to view this object. Even though it has such a low surface brightness, in the right conditions a small telescope is capable of showing this galaxy as a dim unconcentrated glow. A 150mm (6in) aperture makes the glow a little brighter, with a few foreground stars superimposed on the galaxy. Telescopes of 250mm (10in) upwards show a faint, mottled structure over the circular glow. This galaxy is famous for producing supernovae, and in late 2004 it produced one of 12th magnitude, bright enough for a 150mm aperture.

NGC 6960/6992/6995 are parts of a huge bubble of nebulosity called the Veil Nebula, marking the site of an ancient supernova outburst. Distinct parts of the Veil have their own NGC numbers, but only these three are bright enough to see with smaller instruments. NGC 6960 is visible in a dark sky as a long tenuous streak of nebulosity crossing the bright star 52 Cygni; larger telescopes show a conspicuous bend in the nebulosity (Figure 11.10). NGC 6992/95 is brighter and more detailed in telescopes, filling a low-power field of view, but this object really needs a good nebula filter to do it justice; a highly detailed area of nebulosity resembling a long-exposure photograph will then be visible (Figure 11.11).

NGC 7000, the North America Nebula, is a large emission nebulosity shaped like the North American continent. This object is so large that it can be seen with the naked eye from a dark site, as an amorphous glow just west of Deneb. The nebula is two 2° across, so binoculars give a much better view than a telescope.

Delphinus

This small constellation lies close to the Milky Way and contains a few deep-sky objects for the amateur observer.

NGC 6934 is a magnitude 8.9 globular cluster, 7 arc minutes across, visible as a small spot of light in a small telescope. A 150mm (6in) aperture reveals a slightly elongated but bright unresolved glow next to a red magnitude 9.2 star. Even large telescopes at high powers only partially resolve this object, because of its great distance – 62,000 light years.

NGC 6891 is a magnitude 11.7 planetary nebula only 15 arc seconds across. It is a tough object for a small telescope, which will show no more that a stellar spot and remains featureless even in a 150mm (6in) telescope. The central star, of magnitude 12.5, needs an aperture of at least 200mm (8in) to be seen well.

NGC 7006 is a globular cluster which lies 160,000 light years away. Its magnitude is 10.6, so it is only faintly visible in a 150mm (6in) aperture, and even in a 300mm (12in) telescope just a few faint stars are resolved.

Dorado

A far-southern constellation, Dorado contains the Large Magellanic Cloud, an irregular satellite galaxy of the Milky Way, 160,000 light years away.

The **Large Magellanic Cloud** (LMC) is visible to the naked eye as a large, 6° nebulous glow in the south of the constellation. Any optical equipment will show many intriguing areas within the cloud.

NGC 2070, the Tarantula Nebula, is the best object within the LMC. It is a huge, looping gas cloud about 1,000 light years in diameter, and if it were at the same position as the Orion Nebula (M42) it would fill the entire constellation of Orion and cast shadows at night here on Earth! The Tarantula is visible with the naked as a fuzzy star, and thus also known as 30 Doradus. Binoculars show a bright, slightly elongated patch of light; most apertures of telescope show an intricate, looped mass of nebulosity with a small cluster of stars at its centre that do indeed look like the eyes of a celestial spider.

Draco

This is a large constellation which winds around the North Pole Star, and as it is far from the galactic plane it is rich in galaxies, though most are rather faint.

NGC 5907 is a magnitude 11.1 edge-on spiral only faintly visible in a 150mm (6in) aperture as a narrow, 6 arc minute streak of light. Larger

telescopes extend the length to a maximum of 12 arc minutes, and in a 300mm (12in) telescope the equatorial dust lane may be seen in good conditions, but no central bulge is visible.

NGC 6543, also called the Cat's Eye Nebula, is a magnitude 8.8 planetary nebula, 20 arc seconds across and an easy object for a small telescope, visible as a tiny spot of light ESE of a magnitude 9.8 star. Its high surface brightness makes any detail difficult to see, even in larger apertures. A 150mm (6in) aperture will show little detail, but an 200mm (8in) at high power and with a nebula filter begins to reveal distinct inner and outer regions and an overall mottled appearance. The magnitude 11.3 central star is difficult because the brightness of the planetary itself overpowers it.

Eridanus

This huge, winding constellation spanning 60° in declination stretches from near Rigel in Orion into the far southern sky.

NGC 1535, also called Cleopatra's Eye, is a magnitude 9.6 planetary nebula, 21 arc seconds across and one of the best objects of its class for amateur telescopes, visible as a stellar spot in a small telescope. A 150mm (6in) aperture shows a conspicuous circular object with a brighter middle; the magnitude 12.6 central star can be seen at a high power, though it takes a larger aperture to show it easily.

Fornax

A small barren constellation, Fornax contains many galaxies though none are spectacular in amateur instruments.

NGC 1316, also known as the Fornax A radio source, is a magnitude 9.8 barred lenticular galaxy – a type of galaxy with a disk and central bulge but no spiral arms. At 11 arc minutes across it is visible in a small telescope as a small fuzzy area; larger apertures don't show much more, though it will appear brighter, with a stellar nucleus.

NGC 1360 is a bright planetary nebula for a small telescope, visible as a well-defined magnitude 9.6 oval haze elongated NE–SW. A 150mm (6in) aperture shows a large glow, smooth in appearance, and the central star of magnitude 11.3 is very conspicuous.

Gemini

Gemini is a bright constellation containing some wonderful deep-sky objects for all telescopes.

M35 is a bright-naked eye (magnitude 5.6) open cluster, partially resolved in steadily held binoculars. A small telescope resolves about 40–45 stars, while a 150mm (6in) aperture will show perhaps a hundred, including a few red-

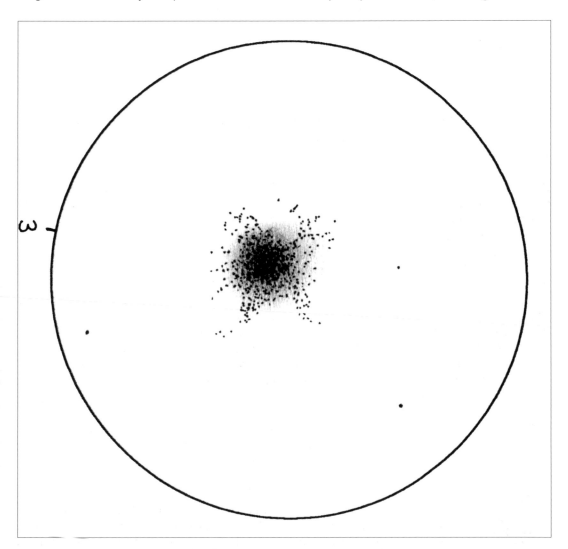

Fig. 11.12 M13, 220mm (8½in) reflector, ×144.

orange members which seem to be arranged in chains. Large telescopes at medium magnification will reveal about 200 stars, filling the eyepiece view.

While observing **M35**, you may notice a small fuzzy area a short distance SW when using a low-power eyepiece. This is NGC 2158, a 12th-magnitude open cluster. It looks very different to M35 because it is much farther away: M35 is 2,800 light years away, whereas NGC 2158 is a much more distant 16,000 light years, making resolution much more difficult. A granular patch is all that a small telescope will show, while a 150mm (6in) aperture reveals about 15–20 faint stars in front of a misty glow. Larger apertures resolve this cluster quite well; it looks more like a globular cluster though with not as many stars.

NGC 2371/2 is a small, faint magnitude 13.0 planetary nebula slightly less than an arc minute across. It is visible in a 150mm (6in) aperture as a hazy patch, while larger instruments will show a double nebula – hence the dual NGC number. The nebula is elongated NE–SW, the SW part is the brighter and larger, and using a high magnification in at least a 300mm (12in) aperture will show the very faint 14th-magnitude central star.

NGC 2392, the Eskimo Nebula, is a magnitude 9.9 planetary nebula easily seen in a small telescope as a fuzzy blue-green star. A 150mm (6in) aperture will show a bright disk with a brighter centre, and the central star, of magnitude 10.5, is easily visible. Larger telescopes show quite a bit of detail within the nebula, including a fainter, diffuse outer area and a distinctly brighter inner region with the appearance of having a coiled structure.

Hercules

Easily recognizable by its distinctive 'Keystone' grouping of stars, Hercules contains the finest globular in the northern hemisphere.

M13, the famous globular cluster, is at magnitude 5.9 visible with the naked eye in good conditions. M13 appears in binoculars as a small, bright fuzzy ball, and a small telescope reveals a very bright ball of grainy light. A 100mm (4in) aperture begins to resolve the extreme outer regions into stardust, and in a 150mm (6in) telescope the cluster is pretty well resolved at high power, but viewed in any aperture over 200mm (8in) it is a sight to behold, with hundreds of stars packed into a 20 arc minute ball and curious dark lanes on the NE side of the cluster (Figure 11.12).

M92 is a smaller, magnitude 6.5 globular cluster, easily visible in a small telescope though no individual stars are resolved. A 150mm (6in) aperture shows a very bright ball of light, with the outer regions resolved only at high magnification, while in a 250mm (10in) instrument the cluster is complete resolved.

NGC 6229, the third globular cluster in northern Hercules, is magnitude 9.4 and faintly visible in a small telescope. In a 150mm (6in) aperture it is visible as a faintly glowing circular area with two 8th-magnitude stars close by; larger telescopes show only a granular patch about 4.5 arc minutes across, and no resolution is evident at any magnification.

Horologium

This far-southern constellation is populated by galaxies, but only one is suitable for amateur instruments.

NGC 1512, a barred spiral of magnitude 11.1, is visible in a 150mm (6in) aperture as a featureless patch of light. Larger telescopes don't do much better, but a faint stellar nucleus is visible at high powers.

Hydra

This is the largest constellation in the entire sky, winding through 7 hours of right ascension, and hosts some fine deep-sky objects.

M48 is a large (54 arc minute) magnitude 5.5 open cluster nicely seen in a small telescope, which will reveal around 60 stars. A 150mm (6in) aperture at low power nicely encompasses the entire cluster, and all the stars appear white.

M68 is a magnitude 8.2 globular cluster, 11 arc minutes across and visible in a small telescope as a dim, circular glow. A 150mm (6in) aperture reveals a few outlying members, while larger telescopes show M68 as a giant ball of light, a 300mm (12in) instrument perhaps resolving the cluster but with some difficulty.

M83 on photographs is a marvellous face-on spiral galaxy of magnitude 7.9, made famous by the first high-resolution image obtained by the Hubble Space Telescope. Unfortunately, amateur instruments do not show much, a small telescope showing only a poorly concentrated, slightly elongated glow. A 150mm (6in) aperture reveals a large misty outer halo and a brighter inner region, while in larger telescopes a brightening is visible on the galaxy's NW side.

NGC 3242 is a bright, magnitude 8.6 planetary nebula, but at less than half an arc minute in diameter it is a tough object in a small instrument, looking simply like a star – but look for the typical blue–green colour that betrays its nature. A 150mm (6in) aperture shows it as a pale blue disk elongated E–W, while larger telescopes show a diffuse outer rim.

Lacerta

This northern constellation is lacking in bright stars but is easy to locate by its distinctive zigzag pattern.

NGC 7209, a magnitude 7.8 open cluster 24 arc minutes across, is visible in a small telescope as a scattering of stars just south of the magnitude 6.2 variable star HT Lacertae. A 150mm (6in) aperture at low power reveals around 50 stars, two of which are red, making a nice colour contrast.

Leo

Leo is a distinctive constellation populated by many faint galaxies, but there are some good examples for the small telescope.

M65 is a bright magnitude 10.2 spiral galaxy, 10 arc minutes long and visible in a small telescope as a bright, elongated patch, while larger telescopes show a granular texture along its length and a dark lane, indicating spiral structure. A 12th-magnitude star is visible just off the NW arm, which is visible in a 150mm (6in) aperture, and a magnitude 13.6 star is visible a short distance to the NE, though you will need a 250mm (10in) telescope to see it easily.

M66 is a magnitude 9.6 spiral galaxy seen nearly edge-on and visible in the same low-power field as M65. It is easily seen in a small telescope as a large elongated patch about 6 arc minutes from the edge of a triangle of 11th-magnitude stars to the NW; a magnitude 9.8 orange star is visible just to the galaxy's NW. Larger apertures reveal a large, bright oval area of light very condensed at the centre, and a magnitude 13.1 star superimposed on the extreme outer southern arm (Figure 11.13).

M95, a bright barred spiral galaxy of magnitude 10.5, is easily visible in a small telescope as a condensed glow. A 150mm (6in) aperture shows a diffuse halo brightening to a concentrated inner region.

M96 is a spiral galaxy visible in the same low-power field as M95. At magnitude 10.1 it is

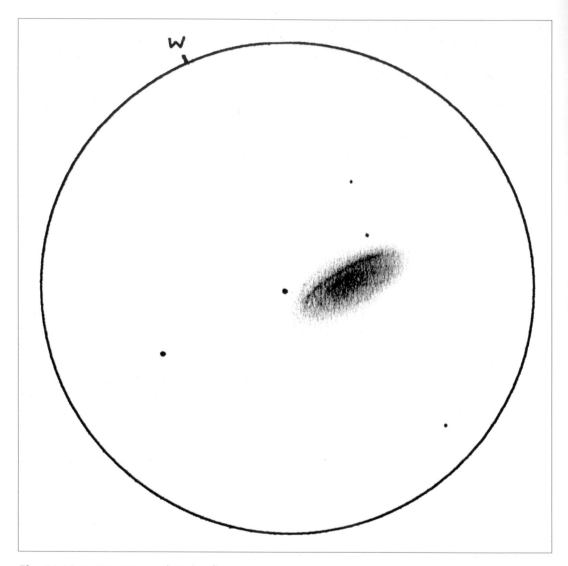

Fig. 11.13 *M66, 300mm (12in) reflector,*
×268.

brighter than M95 in a small telescope and does not look much different in a 150mm (6in) aperture, though larger apertures do show a prominent SE–NW elongation and a sharp stellar nucleus.

M105, an elliptical galaxy within the same low-power field as the M95/96 pairing, is magnitude

10.2, 5 × 4 arc minutes in size, and visible in a small telescope as a hazy area of light. A 150mm (6in) telescope reveals a stellar nucleus, and larger apertures show a diffuse circular patch with a distinctly brighter inner region.

NGC 2903 is bright enough to be included in the Messier catalogue, and it's surprising how Messier did not spot this magnitude 9.6 spiral galaxy. In a small telescope it is visible as a bright, elongated object with a magnitude

10.5 star superimposed on its glow. A 250mm (10in) aperture reveals a bright stellar nucleus, and the galaxy itself appears mottled along its length; look also for a few faint 12th- and 13th-magnitude stars very close to the galaxy's periphery.

Lepus

This small constellation just south of Orion is populated by faint galaxies and also has a rather nice globular.

M79 is a bright, magnitude 7.7 globular cluster which appears as an unresolved haze in a small telescope. It is partially resolved in a 150mm (6in) aperture, and in larger telescopes is nicely resolved.

IC 418 is a magnitude 10.7 planetary nebula just 12 arc seconds across. It is easily visible as a blue-green star in a small telescope, but no disk is present. A 150mm (6in) aperture will show a tiny disk at high power, and the magnitude 10.3 central star becomes visible; in anything larger it is an easy target.

Libra

Libra is a small, faint constellation with many faint galaxies.

NGC 5897 is a magnitude 8.4 globular cluster with a low surface brightness, which makes it a difficult object for a small telescope, and even a 150mm (6in) aperture reveals only a weak haze devoid of any stars. A 250mm (10in) aperture will partially resolve the cluster and show a slight E–W elongation.

Merrill 2–1 is a tiny, 6 arc second planetary nebula of magnitude 11.6, lying west of a magnitude 9.9 star and just east of a magnitude 11.2 star. It is unobservable in a small telescope, and even in a 150mm (6in) aperture it is barely discernible from a star. A large telescope at high power reveals only a tiny, greyish disk.

Lupus

This small southern constellation features a number of globular clusters and planetary nebulae.

NGC 5824 is a bright globular cluster of magnitude 9.1, visible in a small telescope as a condensed glow with a much brighter centre. Larger apertures simply show a much brighter version, and even at high magnifications remains featureless, with no resolution of the cluster's stars.

Lynx

Lynx is a faint, far-northern constellation with many galaxies, though they are not spectacular through amateur instruments.

NGC 2419, the Intergalactic Wanderer, is a magnitude 10.3 globular cluster an amazing 200,000 light years away, making it the most distant of our galaxy's globulars that we know of. Because of its great distance it is a tough object for most telescopes. A small telescope reveals only a barely discernible, low-surface brightness-glow just east of two magnitude 7.5 stars, and even a 300mm (12in) aperture shows no more than a granular appearance (Figure 11.14).

Lyra

Lyra is a small, bright constellation dominated by the bright, blue-white star Vega.

M56 is a small globular cluster, difficult to observe because it lies in the plane of our galaxy, partially obscured by gas and dust. A small telescope will show this magnitude 8.2 globular as a small circular glow in a rich starfield. It appears granular in a 150mm (6in) aperture, and while larger telescopes do resolve it at high power it is still not an easy object.

M57, the Ring Nebula, is the finest example of a planetary nebula in the northern sky, and is

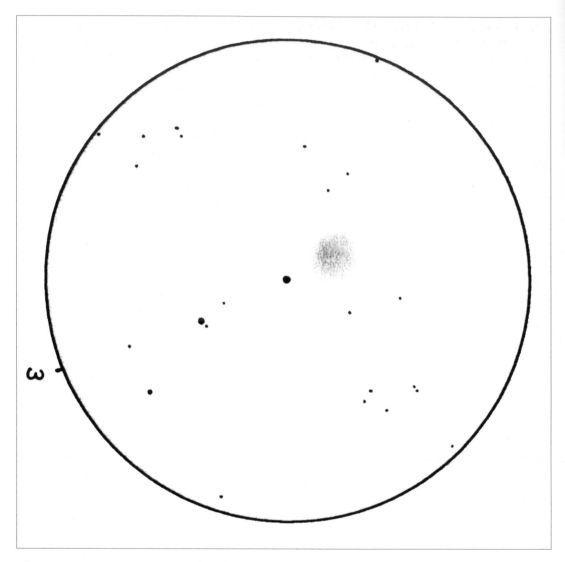

Fig. 11.14 *NGC 2149, 220mm (8½in) reflector, ×208*

visible with a small telescope as a magnitude 9.7 disk 1 arc minute in diameter. A 150mm (6in) aperture at a high power shows a uniformly bright, slightly oval nebulosity with a distinct central hole, and a 12th-magnitude star just off its eastern edge. Larger telescopes show that the central hole in the ring is not completely dark – there is the impression of a very thin veil draped over the object – and the ends of the nebula appear pointed at a high power and fainter than the inner part of the ring. The central star is 15th magnitude, and so will be a very difficult object for large telescopes; it is thought to be variable (Figure 11.15).

NGC 6791 is an open cluster estimated to be 10 billion years old – surely one of the oldest such clusters in the entire sky. A small telescope shows this magnitude 9.5 object as a faint haze, similar to NGC 188 in Cepheus. Large telescopes just begin to resolve the cluster, but the view is still not too impressive. A couple of red and orange stars lie to the east.

Monoceros

Lying in the Milky Way next to Orion, this faint constellation is rich in open clusters and nebulosity.

M50 is a magnitude 7.2 open cluster visible in a small telescope as a scattering of 25 stars, while a 150mm (6in) aperture at a medium power resolves around 80 stars. Larger apertures reveal a total of 100 cluster members, including a red magnitude 7.8 star to the SW; all the other stars are white.

NGC 2237, the Rosette Nebula, so called because of its appearance on long-exposure photographs, is a huge emission nebula 80 arc minutes across visible in binoculars from a dark site under excellent conditions. Through a

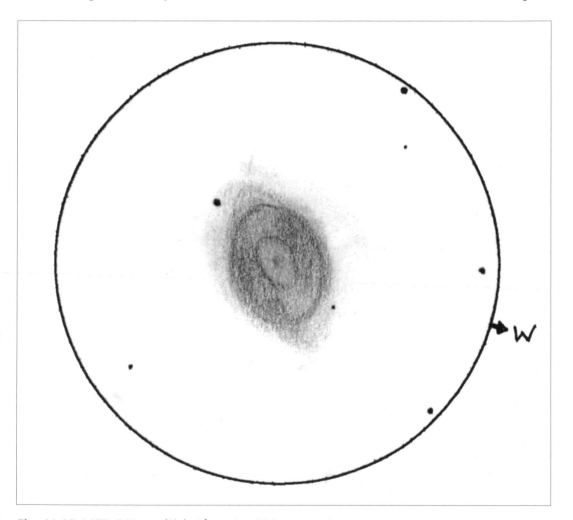

Fig. 11.15 M57, 200mm (8in) refractor, ×333.

telescope this object needs as wide a field of view as possible. If you use a good nebula filter you can expect to see an area of milky-white nebulosity quite condensed in places. At the centre of the nebula is NGC 2247, an open cluster visible in a small telescope as a group of 15 stars.

NGC 2261, Hubble's Variable Nebula, is a small reflection nebula surrounding the variable star R Monocerotis, which illuminates it. It is visible in a small telescope as a small patch of light, but its brightness depends on what stage in its cycle of variability R Monocerotis is at. If the star is near minimum the nebulosity is faint, whereas when the star is at maximum brightness the nebula is an easy object. The nebulosity has a triangular shape; the illuminating star is situated at the tapered end of the nebula, to the south, and at maximum brightness distinct internal mottling is visible running the length of the nebula.

NGC 2264, the Christmas Tree Cluster, is a fine magnitude 4.1 open cluster named for its shape; its brightest star at magnitude 4.6 is 15 Monocerotis. A small telescope will show about 30 stars, and larger apertures show many fainter stars enveloped in faint traces of nebulosity. Superimposed on the faint nebulosity that surrounds NGC 2264 to the south is a small dark nebula called the Cone Nebula. Because of the faintness of the nebulosity against which it is silhouetted, and the Cone's small angular size, it is a difficult object visually.

NGC 2301 is a striking magnitude 6.3 cluster for a small telescope; at a high power a small V-shaped asterism is seen among several fainter stars. Viewed in a 150mm (6in) aperture, the cluster looks rather like M44, the Beehive in Cancer, with an 8th-magnitude red star at its heart.

Ophiuchus

Ophiuchus is a large constellation containing many galaxies and globular clusters, seven of which are Messier objects.

M9 is the constellation's first Messier globular cluster, a magnitude 7.8 object just over 12 arc minutes across and visible as a bright ball of light in a small telescope. A 150mm (6in) aperture shows a granular disk, perhaps just on the verge of resolution, while larger apertures resolve the cluster into countless fine points of light, like white pepper sprinkled on a black surface.

M10 is magnitude 6.6 globular cluster 20 arc minutes across, located 3° SE of M12. A small telescope reveals a bright ball, rather granular in appearance, but a 150mm (6in) aperture resolves the outer regions well, although over a fuzzy halo. Large apertures fully resolve the cluster and show arms stretching from the core to the north and south. M10 is only 1° from the bright multiple star 30 Ophiuchus.

M12, of magnitude 6.1 and 16 arc minutes across, is an easy globular cluster for a small telescope; it appears similar to M10 but has a few magnitude 10–10.5 foreground stars superimposed on it, though the cluster itself is not resolved. A 150mm (6in) aperture at high power partially resolves M12, and larger telescopes completely resolve it.

M14 is a magnitude 7.6 globular cluster, more diffuse than M12 but still an easy target for a small telescope, looking like a slightly elongated ball but with no stars resolved. Even a 150mm (6in) instrument is not sufficient to resolve any stars, and it merely looks granular. M14 in a 300mm (12in) aperture is still a pretty disappointing sight, appearing as a poorly resolved glow.

M19 is another bright globular cluster of

magnitude 6.8, visible in a small telescope as a bright unresolved ball of light; it takes a 150mm (6in) aperture to achieve partial resolution at a high magnification.

M62 is another bright, magnitude 6.4 globular cluster, easy in a small telescope, looking similar to M19. A 150mm (6in) aperture shows a very bright inner core surrounded by a fainter, more diffuse outer halo; there is no resolution at any power, though it does look quite granular. Large telescopes resolve the outer edges, but the core remains granular.

M107, the final globular cluster and Messier object in Ophiuchus, is magnitude 7.8 and 13 arc minutes across. Through a small telescope it is a poorly concentrated glow with no resolution; even a 150mm (6in) instrument shows only a granular appearance, though

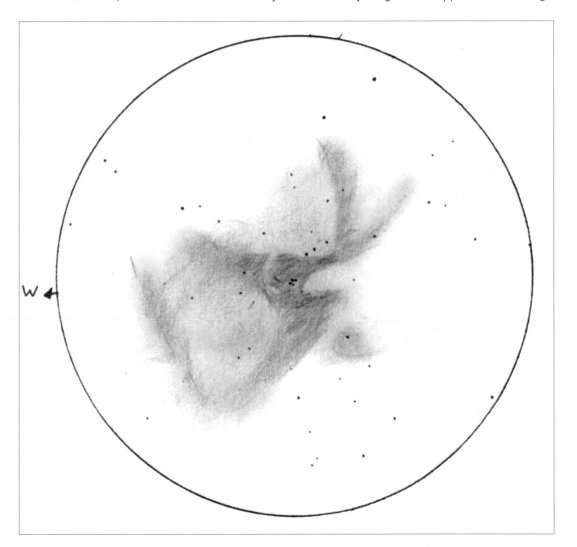

Fig. 11.16 M42, 220mm (8½in) reflector, ×65, UHC, OIII and H-Beta filters.

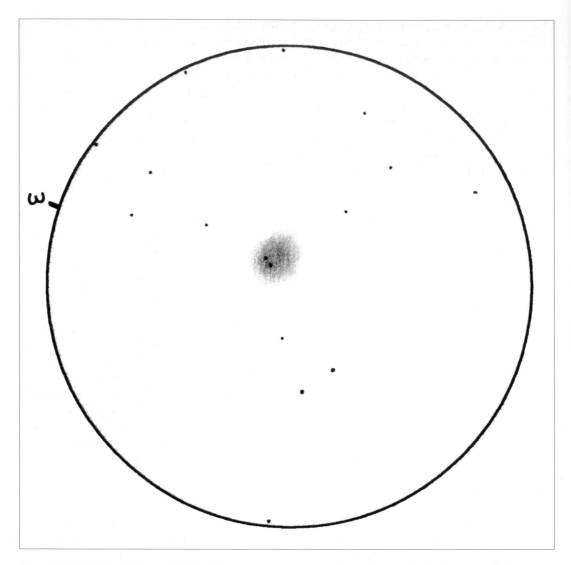

Fig. 11.17 M78, 220mm (8½in) reflector, ×41.

larger apertures do begin to show a little resolution.

Orion

This magnificent constellation, well seen from both the northern and southern hemispheres, contains some of the best of all deep-sky objects for the visual observer.

M42, the Orion Nebula, is one of the most beautiful and most observed objects in the entire sky. It is visible with the unaided eye as a fuzzy star, and binoculars show a small patch of uneven nebulosity centred on a few stars. Small telescopes reveal a large, complex area of bright nebulosity quite extensive to the NW, and the western side of the nebula arcs to the south and SE. Apertures of 150mm (6in) and above show considerable detail within M42; Theta Orionis, the multiple star that illuminates the nebula, is easily resolved into four components – known as the Trapezium – and in good seeing two fainter stars are visible.

In larger apertures the whole nebula has tremendous detail, appearing extremely mottled, especially around the brighter inner region, and in good seeing conditions it can be quite difficult to sketch accurately. Also in large apertures, is the nebula is bright enough to show colour – though only expect subtle pastel shades of grey-green and greyish reds. A conspicuous dark area to the north of the nebula prominent in all apertures is called the Fish's Mouth from its shape and runs SW, stopping about 1 arc minute SW of Theta Orionis (Figure 11.16).

M43 is a small area of nebulosity which is actually part of the M42 complex, but is divided from it by dark nebulosity and appears visually as a separate object. Through all apertures it looks like a backward comma, and has a magnitude 6.8 star at its centre. A 150mm (6in) aperture easily shows a brighter, smooth inner region which larger instruments show as slightly textured.

M78, a small magnitude 8.0 reflection nebula 8 arc minutes across, is easily seen in a small telescope as a grey, circular fuzzy patch with two embedded 10th-magnitude stars. A 150mm (6in) aperture reveals a third, 11th-magnitude star within the nebula, and a high power reveals a fourth, 13th-magnitude star NNE of the main group. Large apertures show one of the stars within the nebulosity to be double, with components of magnitudes 10.3 and 11.5 separated by 2.1 arc seconds (Figure 11.17).

NGC 1662 is a loose, magnitude 8.0 open cluster visible in a small telescope with a prominent magnitude 8.4 orange star at its centre. A 150mm (6in) aperture resolves around 25 stars, and the orange star is more prominent, but it's not a spectacular sight.

NGC 2169, also called the '37 Cluster' because its arrangement of stars resembles the number 37, is of magnitude 7.0 and easily resolved in a small telescope; the '37' shape is very prominent at a high power. A 150mm (6in) aperture will show around 20 stars, of which the brightest is a double of magnitudes 7.4 and 8.0, separation 2.5 arc seconds.

NGC 1981 is a magnitude 4.2 star cluster, visible in the same low-power field as M42, surrounding the magnitude 2.8 star Iota Orionis. A small telescope shows around 15 stars with a pair of equal magnitude 4.8 stars at the cluster's SW edge. Larger telescopes reveal only a few more stars, but show Iota to be a double star with a 7th-magnitude companion 11.3 arc seconds away.

NGC 2022 is a faint planetary nebula located in the hunter's shoulder. With a magnitude of 12.4 and just 19 arc seconds across, it is out of reach of a small telescope, and even a 150mm (6in) aperture shows only a dim, circular patch of light with a few 11th and 12th-magnitude stars encircling the nebula. A 300mm (12in) aperture shows the nebula at its full angular diameter, slightly elongated but with no central star visible.

Pegasus
This constellation, easily recognizable by the four stars of the Great Square, contains many galaxies, though they are too faint for smaller telescopes.

M15, the only Messier object in this huge constellation, is a magnitude 6.4 globular cluster visible in binoculars as a bright fuzzy star. A small telescope shows a bright, condensed hazy ball, though no resolution is evident. A 150mm (6in) aperture reveals a hazy outer halo rising in brightness to a very bright core which at high magnification appears granular. A 250mm (10in) aperture resolves the cluster at high magnification and is a beautiful sight in anything larger.

Perseus

Perseus is partially embedded in the Milky Way and contains mostly open clusters and nebulae.

M34, the first of two Messier objects in the constellation, is a magnitude 5.8 open cluster visible with the naked eye and showing 40 stars in a small telescope, many appearing as double. A 150mm (6in) aperture reveals many doubles among the 50–60 cluster members visible, while larger apertures show around 80 stars, a few with noticeably reddish hues.

M76 is a faint, magnitude 12.2 planetary nebula nicknamed the Little Dumbbell for its resemblance to the Dumbbell Nebula, M27. This small, 1.1 arc minute object is visible in a small telescope as a pale, quite faint patch of light elongated NE–SW. A 150mm (6in) aperture shows that the SW part of the nebula is brighter, while a larger aperture reveals that the seemingly two separate objects are connected by a faint bridge of nebulosity. The central star (magnitude 15.9) is out of reach of amateur-sized telescopes. Low-power views show a very red magnitude 6.7 star (SAO 22551) 12.5 arc minutes to the west (Figure 11.18).

NGC 869/884, also known as the Double Cluster, is one of the northern sky's best deep-sky sights, consisting of two 29 arc minute clusters side by side. They are visible with the naked eye as two misty areas, while binoculars resolve some stars over a misty background. A small telescope shows the two clusters in the same low-power field. NGC 869 appears the more compact of the two, with a magnitude 6.6 star at its centre and a few dozen cluster members scattered around it. NGC 884 is looser, centred on a magnitude 8.8 red star; there is also a prominent 8th-magnitude red star between the two clusters. A 150mm (6in) aperture reveals a multitude of stars in both clusters, plus many more red stars in both.

NGC 1499, the California Nebula, is an extensive emission nebula, named for its resemblance to the outline of the US state. Its integrated magnitude is 5.0, but its extent of 2° gives it a very low surface brightness, making it a difficult object to see. If viewed from a very dark location it is visible in a small telescope as a faint glow near the star Xi Persei. Its size demands a low-power, wide-field eyepiece and as large an aperture as possible, in conjunction with a good nebula filter such as the H-Beta filter from Lumicon.

Pisces

This long, faint constellation contains many galaxies.

M74 is a faint, face-on spiral galaxy with a low surface brightness. This 10 arc minute object is visible as a small, faint patch of light in a small telescope, and a high power reveals a magnitude 12.1 star to the SW. A 150mm (6in) aperture reveals a moderately bright circular glow with about five stars superimposed on the ghostly disk, and larger apertures increase the number of stars to ten. The galaxy itself appears mottled in a 250mm (10in) aperture, and brighter knots are visible here and there across its face; a high power shows M74 to have a bright, condensed core.

Puppis

Puppis stretches to quite a deep southerly latitude and is rich in all types of deep-sky object.

M46 is a remarkable magnitude 6.6 open cluster visible in a small telescope as a rich grouping of 30 faint stars. Within the cluster, towards its northern edge (though in reality a background object 900 light years more

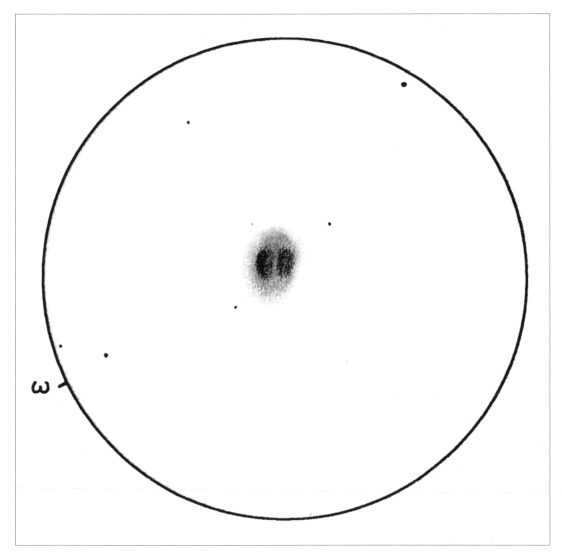

Fig. 11.18 M76, 220mm (8½in) reflector,
×130, OIII filter.

distant) lies a small, 11th-magnitude planetary
nebula (NGC 2348) visible in a small aperture
as a small pale blue disk, easily distinguishable
from the pinpoints of stars that surround it. A
150mm (6in) aperture at a medium power
shows around 70 stars in M46, without any
noticeable concentration; the planetary nebula
is easily seen, though no detail is visible.

M47 is a bright, magnitude 4.3 open cluster,
29 arc minutes across and a pretty sight in a
small telescope, which will show four stars
between 5th and 8th magnitude in a distinctive
keystone shape, with around 30 fainter stars
scattered around them. The brightest star in
the keystone is a double of magnitudes 5.66
and 12.2, with a 5.2 arc second separation. A
150mm (6in) aperture reveals about 50 stars in
the cluster, with a few red stars intermingled
among the predominantly white stars.

111

M93 is a moderately faint, magnitude 6.5 cluster of stars for a small telescope; high magnification reveals about 25 stars, the brightest of which is magnitude 8.2. A 150mm (6in) aperture shows a few dozen stars arranged in a triangular shape, including a few red stars.

NGC 2423 is a magnitude 7.0 open cluster visible in a small telescope as a faint gathering of 15 stars with a 9th-magnitude star at the centre. This, the cluster's brightest star, is a double, with a magnitude 10.2 companion, but its separation is a very close 0.5 arc seconds, requiring a large aperture to resolve. A 150mm (6in) aperture reveals a sparsely populated cluster of around 50 stars, a few with pastel shades of colour.

NGC 2451 is a large, magnitude 3.7 open cluster surrounding the variable star c Puppis. A small telescope will show a few dozen bright

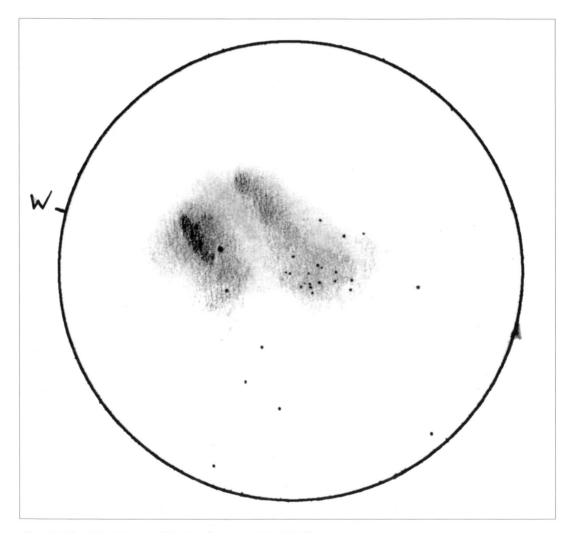

Fig. 11.19 M8, 300mm (12in) reflector, ×84, OIII filter.

stars, concentrated more on the western side; c Puppis is instantly recognizable as a red star, and another red star is visible to the NE of the cluster. The cluster is a beautiful sight in a 150mm (6in) aperture, dozens of stars of all magnitudes filling the field of view; the two red stars have very noticeable hues, and in this aperture some distinctly yellowish stars are seen.

NGC 2477 is a magnitude 5.7 open cluster lying slightly SE of NGC 2451. This 27 arc minute cluster is visible in a small telescope as a gathering of faint stars, but a 150mm (6in) aperture reveals a cluster which looks more globular than open, with over 100 stars packed into a medium-power field of view. Many of the stars have red hues. Large telescopes show an immensely rich cluster, with several hundred stars arranged in clumps and long, arcing chains.

NGC 2546 is a rich, magnitude 5.2 open cluster 40 arc minutes in diameter and consisting of around 50 stars, an impressive sight even in a small telescope. There is a bright magnitude 6.4 star to the SSE, and to the NW a small, curious grouping of 18 stars between magnitudes 8 and 12 that almost looks like a smaller, separate cluster, but is part of NGC 2546. A 150mm (6in) aperture reveals several dozen stars of various magnitudes, including five or six red stars, and larger apertures show a multitude of stars filling the field of view.

Pyxis
This small southern constellation contains only a few objects of interest.

NGC 2627 is a magnitude 8.4 open cluster; 11 arc minutes across and visible in a small telescope at a high power as a group of around a dozen faint stars. A 150mm (6in) aperture reveals about a dozen moderately bright stars, one of which is a 10th-magnitude red star located at the SE of the cluster, and a handful of fainter stars intermingled with them.

NGC 2818 is a magnitude 13.0 open cluster poorly seen in a small telescope as a grouping of just a few stars, while a 150mm (6in) aperture reveals around 15 faint stars, including a planetary nebula nestled within these faint points of light. This planetary, NGC 2818A, is magnitude 11.9 and is visible at high power in a 150mm (6in) aperture as a small disk just over half an arc minute across. Larger apertures show around 80 stars in the cluster, while the planetary is about one arc minute across and its 13th-magnitude central star may be visible.

Sagitta
This is a small, arrow-shaped summer constellation in the northern hemisphere situated between Vulpecula and Aquila. The Milky Way runs through the constellation, which is populated by open clusters.

M71 is a magnitude 8.3 open cluster, 4 arc minutes across and visible in a small telescope as a handful of faint stars at a high magnification; the brightest star in the field is of magnitude 10.4. A 150mm (6in) aperture reveals a few dozen stars over a hazy background, and large apertures reveal about a hundred stars.

Sagittarius
This magnificent constellation contains many wondrous deep-sky objects. It lies in the direction of the centre of our galaxy, with truly awesome Milky Way starfields, but unfortunately for observers at higher northern latitudes Sagittarius never rises very high above the horizon. This constellation contains fifteen Messier objects – the most in any single constellation.

M8, the Lagoon Nebula, is the constellation's grandest deep-sky object, equal in stature to the magnificent M42 in Orion. This large emission nebula is around 1.5° across at its broadest and is easily visible with the naked eye as a bright patch in the Milky Way. A small telescope shows a very bright, condensed nebulosity centred on the magnitude 5.9 star 9 Sagittarii, and a dark area appears to bisecting the nebula, next to the bright open cluster NGC 6530. A 150mm (6in) aperture reveals the nebula to be very bright and condensed, the dark bar is better-defined, and more stars are seen superimposed on the nebulosity.

Larger telescopes show a memorable view: the dark lane has a few faint stars superimposed on it, and it curves to the NE. Many more stars are visible within the nebulosity, especially around 9 Sagittarii. The use of a nebula filter on this object will enhance the view considerably (Figure 11.19); even a small telescope will reveal a vast amount of detail if a UHC-type filter is employed.

M17, the Swan Nebula, a spectacular emission nebula of magnitude 6.0, visible in binoculars as a small fuzzy area, and a small telescope reveals a bright bar of nebulosity orientated NW–SE; the nebulosity seems mottled, even in a small aperture. A 150mm (6in) aperture shows that the nebulosity is shaped like a number 2 – or like a swan, as its name suggests; several stars are seen either superimposed on the nebula or around its periphery. In large apertures, especially when a nebula filter is used, M17 looks like much like it does on a long-exposure photograph: considerable detail can be seen along its length, and a larger but fainter nebulous glow extends beyond the brighter regions (Figure 11.20).

M18 is a small, magnitude 7.5 open cluster visible in a small telescope as a group of 10–15 stars in the shape of an arrow; larger apertures don't improve the view much.

M20, the Trifid Nebula, is a mix of emission and reflection nebulosity surrounding a small group of stars. A small telescope shows a hazy area around a magnitude 7.2 star, with a magnitude 7.3 star to the north. A 150mm (6in) aperture reveals a bright nebulous area peppered with stars, the main magnitude 7.2 star is seen to be double (with a magnitude 10.4 companion 6.2 arc seconds away), and other cluster members are visible; there is also a hint of dark dust lanes crossing the nebula. Large apertures show these dust lanes well against the brighter background nebulosity, and a nebula filter will enhance the view.

M21 is a magnitude 7.2 open cluster visible as a small knot of stars in a small telescope not far from M20; high powers show a small circlet of stars, the brightest of which is magnitude 7.2. A 150mm (6in) aperture shows that the brighter circlet of five stars joins onto a larger and much fainter circlet of eleven stars of between 9th and 12th magnitudes. There is a very red magnitude 10.6 star to the cluster's extreme east.

M22 is a fabulous magnitude 5.1 globular cluster visible with the naked eye as a fuzzy star, while a small telescope reveals a bright ball of light, very fuzzy with minute stars peppered across the haze. A 150mm (6in) aperture shows a huge ball of stars, though the inner region is still fuzzy; in larger apertures arms of stars are seen cascading SE and NE, the outer edges are very irregular, and many red and orange stars are visible.

Owners of large telescope can try to locate the tiny planetary nebula designated PN G009.8–07.5, nestled within the multitude of stars that make up M22. This planetary is of 15th magnitude and only 8.5 arc seconds across, so a steady atmosphere, a large telescope and a high magnification are essential. In addition, try using an OIII nebula filter as this will dim the stars in the cluster while making the planetary nebula more noticeable.

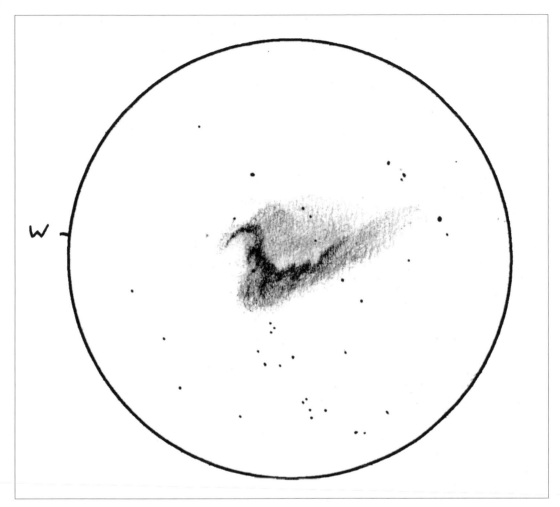

Fig. 11.20 *M17, 300mm (12in) reflector,*
×84, OIII filter.

M23 is a bright, magnitude 5.5 open cluster visible in a small telescope as a sprinkling of 50 stars. A 150mm (6in) aperture reveals around 70 stars, with a couple of red stars dotted here and there; larger telescopes show no more stars, but do show that the cluster members are evenly distributed.

M25 is a bright open cluster for the naked eye of magnitude 6.2, and an easy target in binoculars. A small telescope shows a bright grouping of around 30 stars, including the magnitude 6.6 variable U Sagittarii, which is also a multiple star with many components. A 150mm (6in) aperture reveals 50 stars, many of them red in colour, including many doubles.

M28 is a magnitude 6.9 globular cluster visible as a small fuzzy glow in a small telescope not far from the star Lambda Sagittarii. In a 150mm (6in) aperture it appears brighter, but unresolved; it takes at least 200mm (8in) to achieve partial resolution, but then only at a high power, and a 300mm (12in) aperture resolves the globular but over a misty background.

M54 is a small, magnitude 7.7 globular visible in a small telescope as a fairly bright area. A 150mm (6in) aperture reveals a very bright circular ball of light without resolution, and even in a 300mm (12in) aperture there is no resolution, but it does look granular.

M55 is a bright globular cluster of magnitude 6.3 and a respectable 19 arc minutes in diameter, seen in a small telescope as a large, bright hazy area and slightly mottled. A 150mm (6in) aperture resolves the cluster at a medium magnification, while large apertures fully resolve it.

M69 is another globular cluster, of magnitude 7.7, visible in a small telescope as a small, bright circle of nebulosity with an 8th-magnitude star to the NNW. A 150mm (6in) aperture at high magnification resolves the outer areas; the central part appears as a very bright condensation.

M70 is a small globular cluster visible in a small telescope as a concentrated glow with a few 9th- to 12th-magnitude stars feebly shining over it. A 150mm (6in) aperture shows these probable foreground stars more clearly, but the cluster itself appears granular and is partially resolved only in larger apertures.

M75, the final Messier object in the constellation, is a small, magnitude 8.6 globular cluster, once again seen only as a fuzzy circular glow in a small aperture. A 150mm (6in) telescope reveals a slightly larger glow with a very bright core, though no resolution is evident, but a magnitude 11.1 star is prominent at the cluster's SE edge. Larger apertures only show a granular appearance.

Scorpius
This is a large constellation containing four Messier objects and a large number of open clusters.

M4 is a large globular located a few degrees east from Antares, the constellation's brightest star. M4 is of magnitude 5.4 and is visible to the unaided eye as a fuzzy star, while a small telescope will show a large, smooth weak glow with a few faint 9th- to 11th-magnitude stars peppered across it. A 150mm (6in) aperture partially resolves the cluster; the broad, concentrated inner glow looks granular, and the outer regions are resolved into stardust. Larger telescopes at high magnification resolve M4 into a mass of faint stars.

M6 is a magnificent naked-eye star open cluster of magnitude 4.6 located in the scorpion's 'sting'. Binoculars resolve the cluster, which is centred on the magnitude 6.7 variable star V862 Scorpii, and a small telescope reveals around 50 stars, which seem more concentrated to the east, where the prominent 6th-magnitude red variable star BM Scorpii can be seen. A 150mm (6in) aperture shows many fainter stars scattered among the prominent brighter members, including a red star of magnitude 9.7 a little to the SW of the cluster's brightest member.

M7 is a magnitude 3.3 open cluster easily visible with the unaided eye and is a fine sight in binoculars. A small telescope shows a bright, well-resolved group of sparkling stars, a few with slightly red-orange hues, which almost fills a low-power field of view. A 150mm (6in) aperture reveals countless cluster members in a low-power field, consisting of white, blue-white, red and orange stars.

After the last two Messier entries, **M80** is a bit of a disappointment, but it is easily found halfway between Alpha and Beta Scorpii. This is a magnitude 7.2 globular cluster visible in a small telescope as a small fuzzy blob. A150mm (6in) aperture reveals a moderately bright ball of light, slightly granular in appearance at a high power; it takes at least a 300mm (12in) aperture for full resolution.

NGC 6124 is a bright, magnitude 6.3 open cluster consisting of a few dozen 9th- to 11th-magnitude stars; it is just visible with the unaided eye and a nice sight in a small telescope. A 150mm (6in) aperture at low power reveals about 80 stars, the main grouping lying to the west, and several red stars standing out.

NGC 6231 is a bright open cluster of magnitude 3.4 which appears as a bright knot of light with the unaided eye; a small telescope shows around 20 fairly bright stars, nine of which are between magnitude 5.5 and 8. A 150mm (6in) aperture shows at least a hundred stars going down to threshold magnitudes, including some nice blue-white stars. There are at least six double stars in the cluster, some of which have several components, which are resolvable with a medium aperture.

NGC 6242 is a small, magnitude 8.2 open cluster, 9 arc minutes across and visible in a small telescope as a group of ten moderately faint stars, the brightest of them, of magnitude 7.3, exhibiting a reddish hue. A 150mm (6in) aperture at high power reveals a moderately scattered cluster of 7th- to 13th-magnitude stars; the brightest star is seen to be double, with a magnitude 13.8 companion 12.3 arc seconds away.

Sculptor

Sculptor is an inconspicuous constellation containing no bright stars but some interesting telescopic galaxies.

NGC 55, the brightest member of the Sculptor group of galaxies about 8 million light years distant, is a magnitude 9.6 barred spiral galaxy visible in a small telescope as an elongated NW–SE area with a few 11th-magnitude stars on its periphery. Larger apertures show a huge, elongated object nearly half a degree in extent; it is also quite mottled along its length, with small knots of light here and there.

NGC 253 is a magnificent magnitude 8.2 galaxy, visible in binoculars in good conditions; a small telescope shows a bright spindle elongated NE–SW just north of two stars of magnitudes 8.9 and 9.2. The galaxy is seen to stretch to 25 arc minutes in a 150mm (6in) aperture, though no detail is visible and it appears quite smooth. Larger apertures show extensive mottling along the plane of the galaxy; the overall brightness is quite regular, only the central core appearing slightly brighter, and at high powers a long, dark dust lane is visible to the NW of the core. Several faint stars are visible superimposed on the extensive glow of the galaxy.

NGC 288 is a magnitude 8.1 globular cluster located slightly NNW of a magnitude 8.5 star. It is visible in a small telescope as an unresolved patch of light, and it takes at least a 250mm (10in) aperture and high power to resolve.

NGC 300 is another member of the Sculptor group of galaxies. It has an integrated magnitude of 8.8, but as it is a face-on spiral with a large apparent diameter, 19 × 13 arc minutes, its light is spread over a large area and it has a very low surface brightness. It is therefore a difficult object in a small telescope, but in excellent conditions it is just visible as a very faint glow, with several superimposed 9th- to 11th-magnitude stars. Larger apertures reveal a bright, featureless glow with several stars down to 14th magnitude peppered across its face.

NGC 613 is a magnitude 11.0 spiral, seen nearly face-on and 5 arc minutes wide, located just north of a magnitude 9.9 star. A small telescope shows only a small, dim blob of light, but apertures of 200mm (8in) reveal a bright halo with a brighter centre elongated

NNW–SSE. Slightly larger telescopes resolve a hint of a spiral arm to the south of the core, and also three stars of magnitudes 13.1, 14.0 and 14.1 at the extreme SSE of the galaxy.

NGC 7507 is a magnitude 11.6 galaxy easily visible in a small telescope as a bright ball of light just SE of a magnitude 10.6 star. This is an elliptical galaxy, so most apertures won't show much detail apart from a brighter nucleus, but the larger the aperture, the more intense the core; a 300mm (12in) telescope shows a barely perceivable stellar nucleus at a high power.

Scutum

This is a small Y-shaped constellation a little north of Sagittarius, and lies in the plane of our galaxy.

M11, the Wild Duck Cluster, is a magnificent open cluster of magnitude 6.1; it is so dense that it looks more like a loose globular cluster, and is a fine sight in any optical instrument. A small telescope shows a partially resolved glow at a medium magnification centred on a magnitude 8.5 star. A 150mm (6in) aperture at around ×100 resolves the cluster into a mass of stars. Larger instruments present a truly beautiful sight, the brighter magnitude 8.5 star standing out against several hundred tightly packed stars which go to threshold magnitudes.

M26 is a small, magnitude 9.0 open cluster only 14 arc minutes across visible as a cluster of perhaps five dim stars, the brightest of which is magnitude 9.1. A 150mm (6in) aperture resolves around 25 stars, and larger apertures show that the fainter members in the cluster are arranged in arcs.

Serpens

Serpens is a large constellation, split into two parts by Ophiuchus and populated by a wide range of deep-sky objects, including two Messier objects.

M5, a magnitude 5.7 globular cluster 23 arc minutes across, is easily found as it is just 22 arc minutes NNW of the magnitude 5.1 star 5 Serpentis. It appears unresolved but granular in a small telescope, whereas a 150mm (6in) aperture at a high power resolves M5 into swarms of minute points of light and shows the centre of the cluster to be very bright and condensed; larger instruments fully resolves the cluster.

M16 is an open cluster (NGC 6611) surrounded by the Eagle Nebula (IC 4703), an impressive emission nebula. A small telescope shows around a dozen stars in the cluster, surrounded by a faint, ghostly area of nebulosity if viewed from a dark site. A 150mm (6in) instrument shows about 25 stars, most of them quite faint, and the nebulosity appears as an irregular glow, the northern side having a 'bite' out of it which is quite prominent through a nebula filter.

Palomar 5 is a difficult globular cluster, 78,000 light years distant; of magnitude 11.8 and only just over 3 arc minutes across, it may just be visible in a small telescope as a tiny, very faint spot. A 150mm (6in) aperture shows a weak, hazy area flanked by a magnitude 9.0 star to the SE and a magnitude 10.6 star to the NW. Larger apertures don't improve the view much.

Sextans

A faint constellation on the celestial equator located just south of Leo, Sextans is populated by galaxies.

NGC 3115, the Spindle Galaxy, is the constellation's most impressive object: a magnitude 10.1 elliptical galaxy visible in a small telescope as an elongated haze orientated NE–SW, though no detail is visible. In a 150mm (6in) aperture the galaxy appears to be about 5 arc minutes in length, with a brighter inner region and a magnitude 12.7 star to the SE. Larger apertures show a very smooth, elongated glow rising in brightness to a prominent nucleus.

Taurus

This magnificent constellation, packed with bright stars, hosts two huge naked-eye star clusters and is the site of a famous supernova.

M1, the first item on Charles Messier's list, is the Crab Nebula, the site of a supernova outburst witnessed by Chinese astronomers in the year AD1054. It was given its popular name by Lord Rosse, who observed it in 1848 with his 72-inch (1.8m) reflector and thought it looked like a crab! This magnitude 8.4 patch of light is

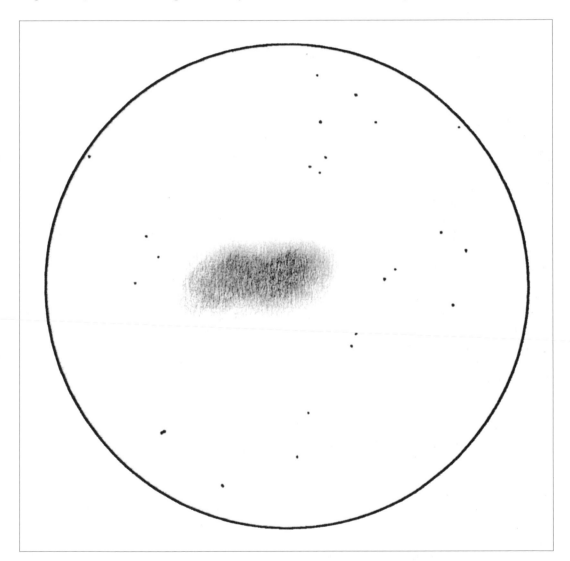

Fig. 11.21 *M1, 220mm (8½in) reflector, ×81, UHC filter.*

visible in a small telescope as a pale patch of light elongated SE–NW just 70 arc minutes NW of the star Zeta Tauri. A 150mm (6in) aperture reveals a brighter area around 5 arc minutes in diameter, and careful viewing may reveal a slight S shape to the nebulosity. In larger instruments this S shape is very evident, and the nebula has a slightly textured look; a few faint stars are seen superimposed on the nebulosity (Figure 11.21).

M45, the Pleiades or Seven Sisters, is so called because the average-sighted person can see seven stars in this open cluster with the naked eye – though from a dark observing site careful viewing can reveal more. This cluster is nearly 2° in diameter, and binoculars give a fabulous view of dozens of blue-white diamonds arranged in a very prominent pattern.

A small telescope shows many dozens of stars in a low-power field, including many doubles, and a low power must be used for the most aesthetically pleasing view. 150mm (6in) instruments reveal more stars, but the field of view is smaller unless an eyepiece with a very wide field is used. Under clear, moonless skies this aperture may show the faint nebula NGC 1435, which is brightest around the star Merope, and known therefore as the Merope Nebula. NGC 1435 is a reflection nebula and will appear as a faint wispy patch of light, more delicate than breath on a mirror; this is only the brightest part of a whole region of nebulosity which totally encompasses the whole cluster and is so evident on photographs.

The Hyades is a very prominent cluster which appears to includes the constellation's brightest star, the magnitude 0.9 orange giant Aldebaran, though this is a line-of-sight effect: the Hyades are at a distance of 150 light years, Aldebaran is only half that distance at 65 light years. The cluster is huge – about 5.5° across, far too large for the widest field in a telescope but perfect for binoculars. A pair of 10 × 50

binoculars will just fit the whole cluster in and show a few dozen white stars, the fiery Aldebaran making a stark contrast. A small telescope will show a few subtly hued orange stars in the cluster, and a 150mm (6in) instrument will reveal quite a few red points of light intermingled with the other blue-white cluster members.

NGC 1514 is a small, magnitude 10.0 planetary nebula only 2 arc minutes across, flanked by a magnitude 8.1 star to the south and a magnitude 8.4 star to the north. A small telescope will barely show this object, as it has quite a low surface brightness, but larger apertures fare much better. A 150mm (6in) instrument reveals a moderately faint circular glow with no apparent detail, though the magnitude 9.4 central star is easily visible, as are few other fainter field stars, and the southern flanking star is seen to be reddish. Larger apertures show a diffuse circular glow with a bright ring around the edge and a bright bar connecting one side to the ring.

NGC 1647 is a large, widespread open cluster of magnitude 6.2 visible in a small telescope as a grouping of 40 stars. There is a prominent double just north of centre, consisting of magnitude 8.7 and 8.9 stars separated by 33 arc seconds, plus two distinctive orange stars of magnitudes 6.0 and 7.5 to the SE. A 150mm (6in) aperture shows about 50 stars with no particular concentration; the two orange stars stand out well, and a few fainter red stars can be seen within the cluster.

NGC 1746 is a magnitude 6.1 open cluster, 41 arc minutes across, visible in a small telescope as a moderately scattered grouping complete containing a few reddish stars, giving a nice colour contrast. A 150mm (6in) aperture at a low power reveals about 80 stars down to threshold magnitudes, including about eight red stars. One of these, on the extreme eastern

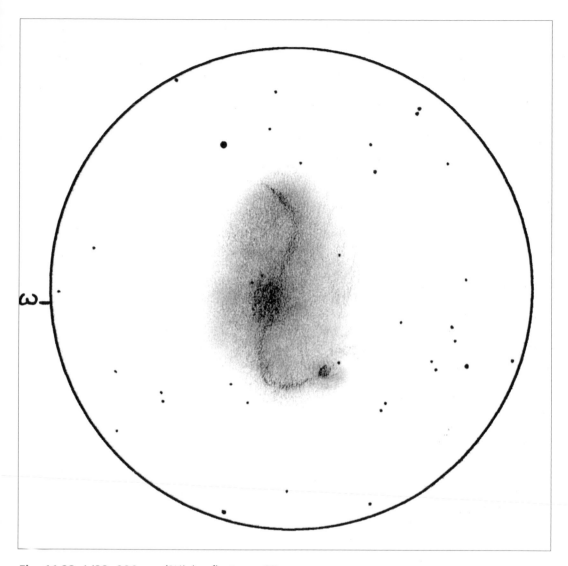

Fig. 11.22 M33, 220mm (8½in) reflector, ×65.

side of the cluster, is double, consisting of a magnitude 7.2 red star with a magnitude 11.9 companion 12.7 arc seconds distant.

Triangulum

Triangulum is a small but easily recognizable constellation rich in galaxies, including M33 – like the Andromeda Galaxy (M31), a member of the Local Group.

M33 is a magnificent, magnitude 5.7 face-on spiral galaxy just over 2 million light years away and just visible with the naked eye in good conditions if you know where to look. A small telescope in good seeing conditions will show a large but weak hazy area orientated NE–SW, and a high power will show that the central region is more condensed, but with no other detail visible. A 150mm (6in) aperture, again in good conditions, shows a hint of spiral structure and also that the central core is much brighter; several stars are visible superimposed on the galaxy, including a magnitude 8.1 star to the north and a red magnitude 8.1 star to the SE.

Larger apertures show M33's total size to be 1° in length. It is very mottled; the spiral structure is quite prominent, appearing like a huge elongated S; and the galaxy is peppered with faint stars. In a 300mm (12in) or larger aperture M33 is a sight to behold. The whole area of milky nebulosity is covered with bright condensations, the brightest of which, located 12 arc minutes NE of the nucleus, is designated NGC 604. This is a huge emission nebula in the outer region of one of the galaxy's spiral arms and is actually visible in a small telescope,

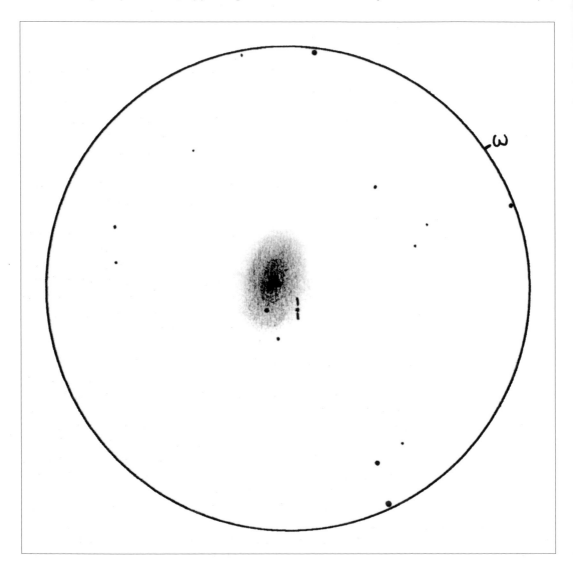

Fig. 11.23 *M81 with Supernova 1993J, 220mm (8½in) reflector, ×104.*

though in a large aperture it looks like a small galaxy in its own right (Figure 11.22).

NGC 672 is a magnitude 11.6 barred spiral galaxy 6 arc minutes across and visible in a small telescope as an elongated haze. Larger apertures reveal the western side of the galaxy to be broader, with a magnitude 13.5 star faintly visible to the NW.

NGC 925 is a magnitude 10.9 nearly face-on spiral galaxy, and like most face-on spirals it has a very low surface brightness, of 14.3 mag/arcmin2. Nevertheless it is visible in a small telescope as a faintly glowing patch of light, elongated slightly NW–SE, but no other detail is apparent. A 150mm (6in) aperture reveals a moderately bright glow with a brighter middle, and also several faint stars in its periphery, including three magnitude 12.7, 12.9 and 13.1 stars embedded in the milky glow. Large apertures show at least eight stars superimposed on the glow of the galaxy, plus a few vague markings along its very mottled length.

Tucana

This small, far-southern constellation is unspectacular in itself but contains some fascinating deep-sky objects for the amateur.

NGC 104, also called 47 Tucanae, is a magnificent magnitude 4.0 globular cluster, 20,000 light years away and visible with the unaided eye as quite a bright fuzzy star – hence its alternative, stellar designation. A small telescope reveals an incredibly bright ball of granular light, the outer edges just beginning to resolve at higher powers. A 150mm (6in) aperture resolves the cluster, though the inner region is merely mottled, apart from a magnitude 8.9 star just south of the cluster's centre and a few brighter stars around the periphery, including a magnitude 8.3 red star at the western side and a magnitude 9.6 red star to the east.

NGC 292, the Small Magellanic Cloud, is a satellite galaxy of the Milky Way lying 195,000 light years away and visible with the naked eye as a nebulous area. Binoculars resolve some of the clusters and nebulosity within the galaxy. Small telescopes reveal a multitude of objects within the glow, and it is seen to be peppered with foreground stars; large apertures show far too much to comprehend – it's best just to sit back and enjoy the view.

Ursa Major

This magnificent northern constellation is instantly recognizable from the familiar grouping known as the Big Dipper or Plough, but the constellation extends far beyond these stars. Ursa Major contains many galaxies within its borders, and there are seven Messier objects – although one is controversial.

M40, although on Messier's list, is simply a double star with the alternative designation Winnecke 4, consisting of two 9th-magnitude stars 50 arc seconds apart and easily resolved in a small telescope. Messier noted the pair when he was searching for a nebula which Johann Hevelius thought he had seen in the vicinity, and although there was no nebulosity associated with the two stars he gave them a number in his catalogue.

M81 is a magnificent spiral galaxy of magnitude 7.8. It is easily visible in binoculars; a small telescope reveals a bright, well-concentrated glow orientated NW–SE a few arc minutes NW of two magnitude 8.6 and 9.8 stars, the fainter of which is triple. M81 itself has diffuse edges and a very bright nucleus, with two magnitude 11.5 stars superimposed near the nucleus.

A 150mm (6in) aperture reveals a very large, oval nebula; it has a very bright inner region that continues down to a brilliant core, and the outer regions are a featureless uniform glow. There are a few stars superimposed on

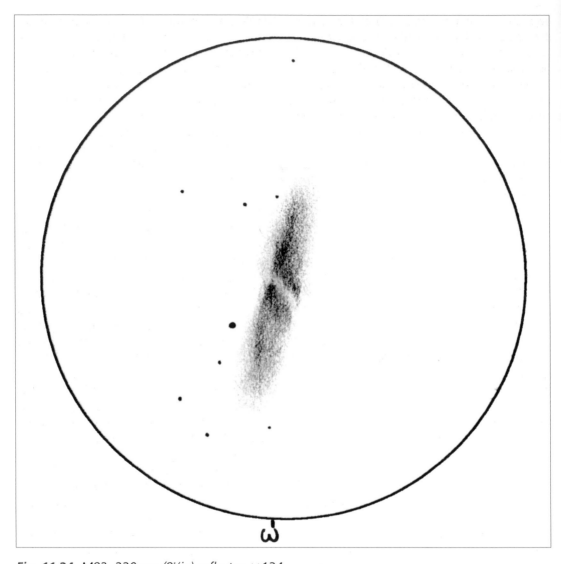

Fig. 11.24 *M82, 220mm (8½in) reflector, ×134.*

the face of M81; the two bright stars to the galaxy's NW are well seen – the magnitude 9.8 triple star is easily split into its three components, and the magnitude 8.6 star at high power is seen to be double, with a magnitude 9.4 companion 2.12 arc seconds away. Large apertures show M81 as a very large halo of light up to 20 arc minutes in extent; the whole length is very mottled at high power, and in good conditions a hint of spiral structure is visible (Figure 11.23).

M82, visible in the same low-power field as M81, is a magnitude 9.2 irregular galaxy. A very prominent dark marking bisecting the galaxy is visible in a small telescope. In a 150mm (6in) aperture M82 is seen as a long streak of light elongated SW–NE; the dark band is easily seen even at low magnifications, and the galaxy itself appears slightly mottled in good seeing. Larger instruments reveal a distinctly mottled silvery streak of light. Two dark areas are visible crossing the width of the galaxy; between these two bands the galaxy is much brighter, and a few bright patches can be seen here and there along its length (Figure 11.24).

M97, the Owl Nebula, is a faint, magnitude 12.0 planetary nebula seen in a small telescope at a high power as a faintly glowing disk of light just NE of a magnitude 11.8 star. A 150mm (6in) aperture reveals a dim but fairly prominent uniform disk, but no detail is discernible; it takes at least a 200mm (8in) aperture to show the famous owl's eyes, but they are still difficult to see unless a nebula filter is used. A 300mm (12in) aperture shows the eyes much better, though they are of poor contrast (Figure 11.25). At this aperture the nebula's edges are quite diffuse, and you may see a mottled texture over the rest of the nebulosity. The central star is of 14th magnitude (though estimates vary quite widely) and can be difficult to view; a 200mm aperture shows it at high power, and it is an easy target for a 300mm.

M101 is a magnificent face-on spiral galaxy of magnitude 7.7. It is visible in a small telescope as a large, dim circular nebulosity with a slightly brighter core; a magnitude 8.1 yellow star is visible very near its northern edge. A 150mm (6in) aperture reveals a much brighter glow with a small but bright nucleus and, in excellent seeing conditions, a slight spiral structure and a magnitude 12.2 star superimposed on the centre of the galaxy. Larger apertures show a bright, mottled haze and a set of spiral arms winding out from a small, bright nucleus; within these arms are many brighter areas with their own NGC numbers. There are several faint stellar points superimposed on the galaxy itself, the faintest of around magnitude 13.5.

M108 is a magnitude 10.9 spiral galaxy visible in the same low-power field as M97 and visible in a small telescope as a very faint patch of light not far from a magnitude 8.9 star to its west. A 150mm (6in) aperture reveals a smooth area elongated E–W. There doesn't seem to be any central brightening at any magnification, and even in large apertures only a subtle central condensation is evident; the galaxy does seem very mottled, even 'lumpy', which may be spiral structure just on the verge of visibility.

M109 is a magnitude 10.8 barred spiral galaxy, of low surface brightness (13.6 mag/arcmin2) but easy to find as it lies only 39 arc minutes SE of the star Gamma Ursae Majoris. In a small telescope it is a difficult object, visible only as a faint, totally smooth patch of light with no visibly brighter centre. A 150mm (6in) aperture shows a faint, diffuse halo, and overall the galaxy looks smooth, though still with no discernible central brightening; a magnitude 13.8 star is superimposed slightly north of centre. Larger apertures reveal a mottled, slightly elongated area with a slight central brightening; the area seems to sparkle, as if there were many extremely faint stars situated here.

NGC 3675 is a bright, highly inclined spiral galaxy of magnitude 11.0 visible in a small telescope as a small, faint patch of light about 5 arc minutes across and elongated N–S. Through a 150mm (6in) aperture it looks very smooth; a bright inner core is visible within the diffuse outer halo, and an almost stellar nucleus can be made out. Also in this aperture a few stars are visible on or around the galaxy, including a magnitude 12.9 star at its northern tip and a magnitude 13.9 star very near its western edge. Large apertures reveal a dark area at the east side of the core; this side of the galaxy seems fainter than the western side.

NGC 3726 is a spectacular magnitude 10.9 spiral galaxy visible as a fuzzy area in a small

125

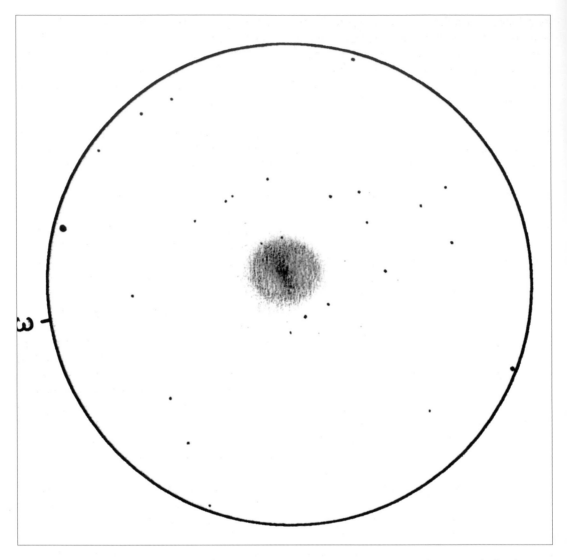

Fig. 11.25 M97, 300mm (12in) reflector, ×84, OIII filter.

telescope. The view is not much improved in a 150mm (6in) aperture, though a magnitude 12.9 star is visible at the NNE tip of the galaxy. In larger apertures the galaxy is seen as a very large, diffuse halo of light, and at high powers the surface is quite mottled, light and dark markings across the uneven glow hinting at a spiral structure.

NGC 3877 is a magnitude 11.8 spiral galaxy 17 arc minutes south of the orange magnitude 3.7 star Chi Ursae Majoris. It is seen as a very faint, elongated patch of light in a small telescope; a 150mm (6in) does not show much more, apart from the NE–SW elongation being more obvious. Large apertures show a fairly bright streak of silvery light which looks granular at high powers, and the core bulges slightly.

NGC 4026 is a galaxy visible in small telescopes

as a long, thin magnitude 11.7 spindle 7 arc minutes SSW of a 9th-magnitude star, oriented N–S. Larger apertures show little more, other than that the nucleus appears brighter and bulges slightly. A few very faint 14th-magnitude stars lie around the galaxy.

Vela

Vela is a mid-southern constellation not easily visible for many northern-hemisphere observers.

NGC 2547 is a magnitude 5.0 open cluster 20 arc minutes across visible in a small telescope simply as a very sparse scattering of stars. A 150mm (6in) aperture resolves only ten or so stars, the brightest of which is magnitude 6.5, and large apertures reveal merely a poor concentration of stars down to 11th magnitude.

NGC 3132, also called the Eight Burst Nebula, is a magnitude 8.2 planetary nebula 30 arc seconds across. In a small telescope it is just visible as a tiny disk, but its magnitude 10.1 central star is more prominent than the nebula. A 150mm (6in) telescope at high power reveals a quite smooth, moderately bright disk of light, the central star once again totally dominating the nebula. Larger apertures reveal a ring-like structure with a brighter area to the SW edge.

NGC 3201 is a bright, magnitude 6.9 globular cluster, 20 arc minutes across and visible in a small telescope as a bright but unresolved ball of light located in a rich starfield. In a 150mm (6in) aperture the cluster's edges are just resolved into stardust, and the centre of the cluster appears very bright and perhaps granular; large apertures resolve several dozen stars over a very bright but very hazy, grainy ball of light.

Virgo

Virgo is a huge constellation that crosses the celestial equator and holds the most observed grouping of galaxies in the entire sky: the Virgo Cluster, with over 3,000 members. Spiral and elliptical galaxies dominate, and some are the finest in their class for the visual observer. All the galaxies listed below are members of the Virgo Cluster.

There are **143 NGC** objects (including a dozen Messier objects) here brighter than magnitude 12.6 – a magnitude figure that should be easily achievable for a 150mm (6in) aperture under a dark sky. For a 200mm (8in) or larger aperture under a dark sky the number of target galaxies, down to a magnitude of 14.0, climbs to over 400. The Virgo Cluster is so dense in places that a 150mm (6in) telescope can have six galaxies in a low-power field of view, and a 200mm (8in) seven or eight, while apertures of 300mm (12in) and larger will show many more.

M84 is a magnitude 10.1 elliptical galaxy 55 million light years distant and over 250,000 light years in diameter. A 150mm (6in) aperture reveals a bright oval haze with a more intense core; larger instruments only make the galaxy look brighter. Also, a 150mm telescope at medium power should pick up a few smaller galaxies dotted around M84 and M86.

M86 is a magnitude 9.3 elliptical galaxy, 10 arc minutes across and easily seen in a small telescope as a bright, elongated spot. Viewed in a 150mm (6in) aperture, M86 is bright with a very intense nucleus, but overall the galaxy appears smooth, as it does in larger apertures.

M87, 60 million light years distant, is a giant elliptical galaxy that lies at the heart of the Virgo Cluster. This is one of the most remarkable objects in the universe: it is one of the largest galaxies known, with a mass eight times that of our own Milky Way, and is surrounded by 4,000 globular clusters. M87 is a strong radio source, and has the alternative

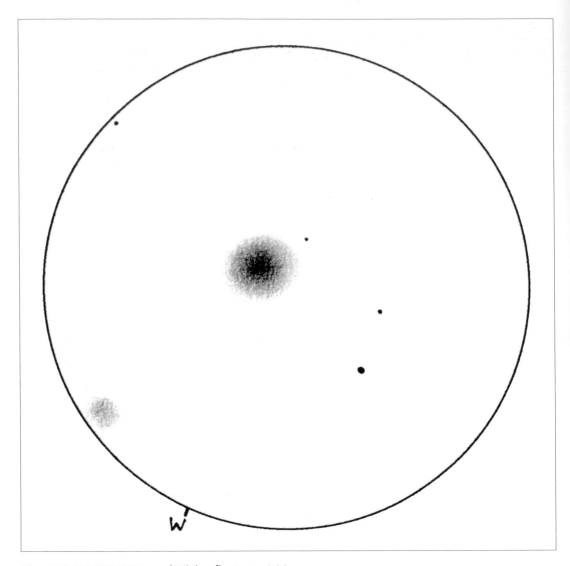

Fig. 11.26 M87, 220mm (8½in) reflector, ×144.

designation Virgo A. In 1918 astronomers at the Lick Observatory discovered a jet of material more than 500,000 light years in length being ejected from the centre of the galaxy.

At magnitude 9.5 this is an easy enough target for a small telescope, but however interesting it is to read about, visually it is a disappointing sight. A high power on a small telescope shows only a fuzzy ball of light, and even large telescopes show little detail – a trait of elliptical galaxies (Figure 11.26).

M104 is called the Sombrero Galaxy as it really does look like a Mexican hat. This spiral galaxy is a magnitude 9.1 object 8 × 4 arc minutes across and easily seen in a small telescope as a concentrated patch of light next to a magnitude 9.7 star. A 150mm (6in) aperture reveals a dark lane that cuts across the southern half of the galaxy and is easiest to see against the bright nucleus; larger telescopes show this dark lane cutting clean through the whole length of the galaxy.

NGC 4378 is a magnitude 12.3 elliptical galaxy, 3.3 arc minutes across, which can be seen faintly in a 150mm (6in) aperture forming an obtuse triangle with two 9th-magnitude stars. A 200mm (8in) telescope reveals a moderately bright object with a stellar nucleus.

Vulpecula

This inconspicuous little constellation lies in the plane of the Milky Way and contains many open clusters and planetary nebulae.

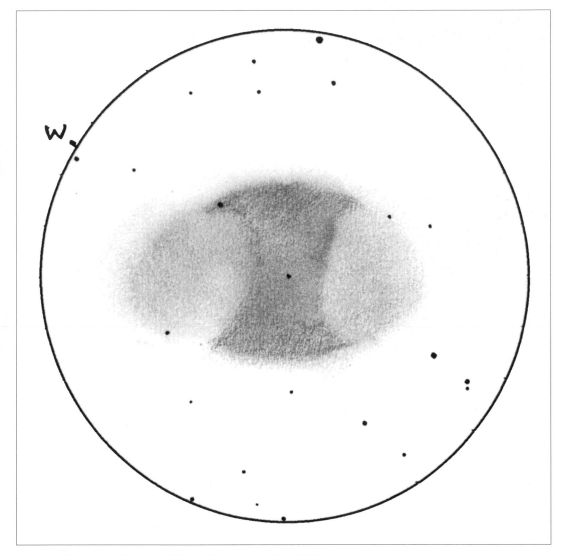

Fig. 11.27 M27, 220mm (8½in) reflector, ×144, OIII filter.

M27, the Dumbbell Nebula, is one of the northern sky's best planetary nebulae. At magnitude 7.6 and 8 arc minutes it is across visible in a small telescope as a rectangular haze with a magnitude 9.7 star on its western edge. A 150mm (6in) aperture reveals a large, bright rectangular nebula. The SW side is brighter, and careful viewing shows a faint Z shape running across and then diagonally down the nebula; also there are a few stars superimposed on the nebula itself. Large apertures show very complex nebulosity, very mottled with brighter areas running along its length; the magnitude 13.9 central star is visible but not prominent (Figure 11.27).

NGC 6940 is a magnitude 7.2 open cluster just over half a degree across and visible in a small telescope as a large, fairly scattered group of around 30 stars. The brightest is of magnitude 8.3 and is at the eastern side of the cluster; and a subtly hued red star of magnitude 9.1 is visible at the centre. A 150mm (6in) aperture resolves around 75 stars. The red central star is much more prominent, and a few much fainter red stars are visible a little to the SSW; the cluster's brightest star is an optical double, with a magnitude 13.3 companion.

Collinder 399 or the Coathanger, so called as it resembles an inverted coathanger, is a bright asterism of magnitude 4.8 and about a degree across. It is an excellent sight in binoculars as it lies in a very rich starfield, and a small telescope, with its wide field, will give a good view at a low power. The cluster is composed of ten main stars, of which 4, 5 and 7 Vulpeculae are the brightest, and around them lie several fainter members. Larger apertures don't usually give good views because their field of view is too small, unless a very-wide-angle eyepiece is used. One of the brighter main stars at the cluster's SSE corner is very orange in colour, while all the other members appear white.

12 Double Stars

Many of the stars visible in the night sky are either double or multiple, and even a small astronomical telescope is capable of resolving hundreds of these often beautifully coloured combinations. Some double stars are sufficiently wide apart that they are easily seen as a pair with the naked eye, for example the wide double Mizar and Alcor in Ursa Major. Other doubles are so close that they are barely resolved (or 'split') in the largest telescopes.

Types of Double Star

There are two kinds of double star, known as optical and physical (Figure 12.1). Optical doubles have no real association with each other: they lie at different distances, yet because they lie on much the same line of sight as seen from the Earth they appear to be close together. Physical doubles are binary stars – two stars gravitationally bound and

orbiting each other in a period that for some systems is measured in months, while others may take tens, hundreds or even thousands of years to complete an orbit. A binary system with more than two members is known as a multiple star.

Mizar and Alcor form probably the best-known optical double. The pair are separated by 12 arc minutes and are easily split with the naked eye. Mizar is 78 light years away and Alcor 81 light years – only 3 light years apart, but not close enough to be gravitationally associated. However, a telescope shows that

Fig. 12.1 *The two stars in an optical double appear close because they lie on the same line of sight for an observer on the Earth. In a physical double the two stars are close in space and orbit each other. The two types of double can look the same through a telescope.*

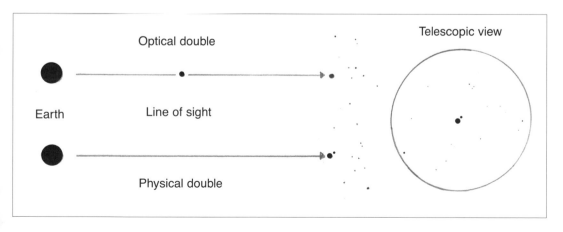

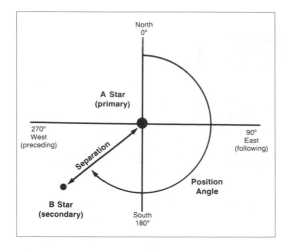

Fig. 12.2 *Determining the position angle of a double star.*

Mizar has another much closer companion and is itself a true physical double – a binary star.

Binary stars are further classified according to the way in which their binary nature has been determined. Visual binaries are wide enough apart for their components to be detected in a telescope. But some are too close to be resolved even in large telescopes. Spectroscopic binaries are recognizable as being double only from periodic changes in their spectra; they can take from a few days to a few weeks to complete an orbit. Astrometric binaries are systems in which one component is much fainter than the other; their binary nature is revealed by a slight oscillation in the motion of the primary star, caused by the gravitational pull of its small companion.

It is interesting to observe a visual binary system with a relatively short period at intervals of several years, as the orbital motion of the two stars will be plain to see. If the orbital period is not too long, it may be possible to observe the changing positions of the two stars over the course of a whole orbit. One such binary is Krüger 60 in Cepheus, a pair of magnitude 9.6 and 11.4 red dwarfs separated (in 2004) by 2.6 arc seconds. Its period is 44.7 years, so it has been possible to calculate its complete orbit, from which predictions can be made of how the two stars will appear in the telescope in future years. The table below gives the year, position angle in degrees and separation in arc seconds (see below) for the next complete orbit of Krüger 60 at five-year intervals.

Krüger 60

Year	PA	Sep.
2004	70.7	2.57
2009	29.3	1.90
2014	308.0	1.40
2019	220.7	1.84
2024	178.2	2.58
2029	153.0	.08
2034	133.4	3.32
2039	115.0	3.30
2044	94.9	3.03

Observing Double Stars

When observing a double star you will need some way of recording which star is which and their relative positions. Astronomers use a system of referring to the stars by letters: the primary (brightest) star is designated A, and the secondary star B; any further components in a multiple system are labelled C, D, and so on. All catalogue entries for double stars list individual stars in this manner.

The orientation of the stars is indicated by the position angle (PA) of the system – the angle between the primary and secondary stars measured clockwise from north. If, for example, star A has a PA of 270°, it is due west of star A (Figure 12.2). Position angles of other components of a multiple star are similarly defined with respect to star A. The apparent

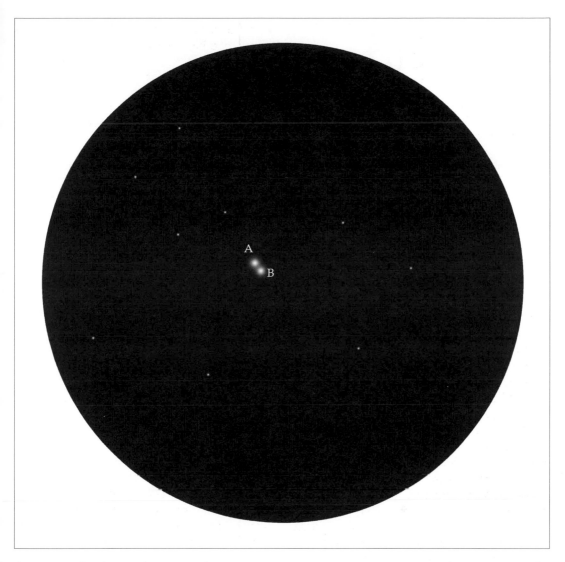

Fig. 12.3 *Gamma Arietis: plain stars can still be beautiful.*

distances of other components from star A are called their separations, and are stated in arc seconds (or, for wider components, arc minutes).

Dawes Limit

In order to know whether a double or multiple star will be resolved in a certain aperture, you first need to look up the separation of the stars in a catalogue. The closer the pair, the larger the telescope you need to cleanly split the stars;

if the aperture of the telescope is too small then the stars will not be resolvable and may appear as a single elongated image.

The nineteenth-century William Dawes determined that a telescope x inches in aperture should just be able to resolve a pair of 6th-magnitude stars separated by 4.56/x arc seconds. This formula is called the Dawes limit,

133

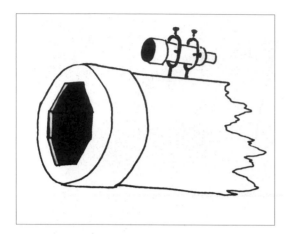

Fig. 12.4 *A hexagonal mask placed over the telescope aperture.*

and in practice it means that a telescope with a 1in aperture will resolve a double star with a separation of 4.56 arc seconds (and an 8in telescope will split a double 0.57 arc seconds apart). For apertures measured in millimetres, the formula is 116/x. To achieve this theoretical limit, the telescope's optics must be of near-perfect quality and the seeing must be excellent.

The Colours of Doubles

Many double stars have beautiful colour contrasts. For example, Beta Cygni (Albireo), the head of the swan, is composed of a golden-yellow primary and a blue-purple companion, and is resolved in 10 × 50 binoculars. Other systems may have no colour but are still visually appealing; an example is Gamma Arietis, often called the 'cat's eyes' because it consists of two pure-white stars, both magnitude 4.8, separated by 16.7 arc seconds – an easy telescopic target (Figure 12.3).

Take particular note of star colours, especially in close doubles, because reported star colours often do not match the observed colour. This is usually an optical illusion: for example, if a red star is next to a white star,

the white star will appear bluish. The individual's eyesight plays a part too – three different observers can observe the same double star and report slightly different colours. In his 1859 book *Celestial Objects for Common Telescopes*, the famous amateur observer the Rev. T.W. Webb described the colours of 95 Herculis, a fine visual double star of magnitudes 4.3 and 5.3 with a separation of 6.4 arc seconds, as apple green and cherry red When I observed it I could only see white and pale yellow! Whatever colours you see at the telescope, write them down, and stick to your judgement – even if you know that other observers have reported different shades.

Double stars are also excellent objects for showing the general public at star parties. Any non-astronomer who looks through a telescope for the first time might not be thrilled by the sight of a galaxy which appears as a barely visible smudge of light, even if you do enthusiastically tell them that it is 90 million light years away. If instead you show them a colourful double star with, say, a golden-yellow primary and a blue companion, you have a much better chance of grabbing their attention and perhaps encouraging them to take up an interest in the subject.

Observing Techniques

Once you have selected a suitable double star, always examine it first at a low power. Unless it is particularly close pair or has a large magnitude difference, you will see some sort of resolution and perhaps a lower power will give a more aesthetically pleasing view. Beta Cygni has a separation of 35 arc seconds, and is just wide enough to be split in 10 × 50 binoculars, and a magnification of ×50 will show the stars to their best advantage, but if a higher power is used the two stars will appear so far apart that the effect of a double is lost.

If the two stars are quite different in magnitude, the use of a high power may help

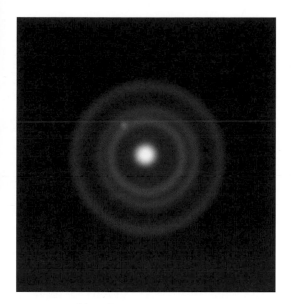

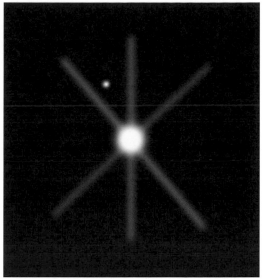

Fig. 12.5 (a) Diffraction rings around a close or unequally bright double can hide the companion star. (b) With a hexagonal mask in place, the diffraction pattern changes to spikes.

to distinguish the fainter companion from the glare of the parent star. Alternatively, there are a few observing tricks to help split a troublesome double. One is to use a high power then move the brighter parent star out of the field of view, thus reducing its glare; the companion star may then become visible, but you must use a high-quality eyepiece with minimal edge-of-field aberrations. Another popular method is to physically hide the brighter star by using an occulting bar – a very thin strip of rigid material, such as a needle, cut to the correct length and glued into position inside the eyepiece – which ensures that the bar is in sharp focus when you look through the eyepiece. Observers usually make their own occulting bar using something that's small enough not to block too much of the field of view when you look into the eyepiece, and opaque so as to fully block the light from the blanked-off star.

A much simpler method is to use a hexagonal mask, which is placed over the aperture of the telescope (Figure 12.4). The perfect image of a star in a telescope consists of a very small disk, called an Airy disk, surrounded by a set of faint diffraction rings (Figure 12.5a). If the companion star is very close to the parent or is much fainter, these diffraction rings may obscure the fainter star. If a hexagonal mask is used, the diffraction rings mimic the shape of the mask – they become six diffraction spikes (Figure 12.5b). The observer simply needs to rotate the mask until the companion star appears between two of these spikes.

Double stars also make very good tests for the quality of your optics. Systems whose separations are between 1 and 2 arc seconds are good tests for telescopes of 100mm (4in) aperture or smaller, and stars with a 1 arc second separation are good targets for a 200mm (8in) aperture. It is not just the separation of the stars that determines whether they can be resolved with any particular telescope: successful double-star observation requires a steady atmosphere, and on misty nights or nights on which the stars twinkle strongly, doubles are not good targets.

The rest of this chapter presents a selection of double stars in the northern and southern hemispheres. Some are easy targets for the smallest telescope, while others will challenge a large aperture and the observing skills of experienced observers. The coordinates, magnitudes, separations and position angles of all these doubles listed here are given in a table at the end of the book.

Northern Double Stars, +90° to 0°

36 Andromedae is a close binary and a test for large amateur telescopes; its components vary in separation from 1.4 to only 0.6 arc seconds over a 165-year period. The primary is a subgiant several times more luminous than the Sun. The companion is probably similar in spectral type because it has the same colour, a lovely brilliant yellow. This system is around 390 light years away, and the 0.6 arc second minimum separation corresponds to the distance between the Sun and Pluto. The two stars are magnitudes 5.5 and 6.6 and at present (2004) are separated by 1.0 arc second at PA 321°, which is a test for a 150mm (6in) aperture, but this will become slightly easier over the coming years as the orbit widens, and in 2034–2039 the two stars will be 1.3 arc seconds apart.

Gamma Arietis is a magnificent pair, bright and easily found. It was discovered to be double accidentally by Robert Hooke in 1664 while he was following a comet. The two stars, of magnitudes 3.9 and 4.6, are separated by a comfortable 7.6 arc seconds, easy for a small telescope. They are both pure white, giving the pair the nickname of the Cat's Eyes.

Eta Cassiopeiae was discovered by Sir William Herschel in August 1779. The primary is magnitude 3.5, and is separated from its magnitude 7.4 companion by 13 arc seconds.

The pair has a most appealing (and controversial) colour contrast: some observers have seen the components as gold and purple, some as yellow and red, and others as 'topaz and garnet'.

Gamma Andromedae is a magnificent double star for small telescopes and a system worth observing at different magnifications. I have observed star B as emerald green at ×65, blue-purple at ×104, then deep blue at ×288. Star A is magnitude 2.1 and the secondary is 5.0, separated by 9.6 arc seconds. A high magnification on a large aperture telescope will reveal that star B is itself a close binary; these two stars are at present (2004) a mere 0.3 arc seconds apart and will continue to close until 2013, when they will be just 0.02 arc seconds apart – far too close for any amateur instrument to resolve. They will be at their widest in 2043, a still very difficult 0.6 arc second separation.

Iota Cassiopeiae is a superb triple star. The primary is magnitude 4.5, star B is magnitude 7.1 and 0.1 arc seconds away, too close for amateur instruments to resolve; the two stars orbit each other with a period of around a thousand years. Star C, magnitude 8.5, is 2.6 arc seconds distant. A and C can be split in a good 75mm (3in) telescope when seeing conditions permit. The primary appears yellowish and the companions are usually described as bluish, but these colours, like those of many double stars, may be illusory. There is also a magnitude 8.9 component (D), 207 arc seconds away at PA 58°.

Alpha Ursae Minoris, or Polaris, the North Pole Star, is a wide optical double with a separation of 18.4 arc seconds. The magnitude difference between the two stars is great, Polaris being magnitude 2.0 and its companion magnitude 9.1, so in smaller apertures the glare from Polaris can obliterate its companion. There are other members of the system, but all are faint and have very large separations.

7 Tauri is a close double and a test for a 150mm (6in) aperture. Star A, magnitude 5.9, and star B, 6.8, are at present (2004) only 0.8 arc seconds apart. This is a binary system with an orbital period of 600 years, and the widest separation, 1.0 arc second, will occur in the year 2322. The closest the two stars will get is 0.2 arc seconds, between 2508 and 2517.

Alpha Tauri, Aldebaran, is the brightest star in Taurus and is a spectacular magnitude 0.9 red giant star 65 light years distant. Aldebaran is an optical double, the companion shining dimly at magnitude 13.7, 30 arc seconds away; such a

Fig. 12.6 Alpha Geminorum.

large magnitude difference makes a positive sighting in smaller telescopes difficult.

Alpha Geminorum, Castor, is an easy physical double for smaller instruments and consists of magnitude 1.6 and 3.0 stars currently (2004) separated by 4.3 arc seconds. Castor has an orbital period of 467 years, the stars having reached their closest in 1969; they will be at their widest in 2083, with a separation of 7.4 arc seconds. A third, fainter star, C, is of magnitude 8.8 and lies 72.5 arc seconds away (Figure 12.6).

Castor is also a spectroscopic binary: each of its three visible components has a companion too close to be resolved through a telescope – three sets of secret twins disclosed only by their spectra. Star C is an eclipsing binary, a pair of nearly matched red dwarfs undergoing mutual eclipses. Two eclipses take place during each orbital period of 19.5 hours, and on each occasion the stars' combined light is halved, dropping by about 0.7 magnitudes.

Zeta Cancri is a binary system consisting of a magnitude 4.7 primary (A) and a magnitude 6.1 secondary (B). These two stars have an orbital period of 59.5 years and at present (2004) are separated by 1.0 arc second, widening to a maximum of 1.2 arc seconds in 2016, which will be a test for an 200mm (8in) aperture. Star C of the system is magnitude 6.2 and is presently 5.9 arc seconds from star A, and with its orbital period of 1,115 years it will stay at this distance for some considerable time. Maximum separation will occur in 2551, when the stars will be 9.5 arc seconds apart.

Gamma Leonis, an easy double for small instruments, consists of a magnitude 2.0 primary and a magnitude 3.6 secondary, currently (2004) 4.4 arc seconds apart. I see both stars as yellow, although other observers have reported different colours. This system has

an orbital period of 618 years and the widest separation, 4.6 arc seconds, will be between the years 2055 and 2071.

Iota Leonis is a fast-moving binary consisting of a magnitude 4.0 primary and a magnitude 6.9 secondary separated by 1.9 arc seconds. The widest separation, 2.7 arc seconds, will be in 2053 – a test for a 150mm (6in) aperture.

Mizar and **Alcor** are the northern sky's easiest visual double, and easy to find, being the stars that form the bend in the handle of the Plough (Big Dipper). Mizar is magnitude 2.2, and Alcor (designated star D) lies 12 arc minutes distant (Figure 12.7). This large separation makes the two stars an easy target for the naked eye, and of course they are seen excellently in even the smallest binoculars. A small telescope reveals that Mizar itself is double; its companion, designated Star C, is 14.4 arc seconds distant at PA 152°. A and C form a true binary system.

Epsilon Boötis is one of the most beautiful double systems in the northern sky, consisting of a golden-yellow magnitude 2.3 primary and a blue magnitude 4.8 secondary 2.9 arc seconds away (Figure 12.8). It is a difficult object in smaller telescopes, and in bad seeing conditions is difficult in a 150mm (6in) telescope.

Mu Boötis consists of a magnitude 4.3 pale yellow primary and a yellowish magnitude 7.2 secondary. The two stars are 108 arc seconds apart, easy in all apertures, but look closely at the fainter star with a 150mm (6in) aperture at high magnification and you will find that it itself is double, with a fainter magnitude 7.7 star a mere 2.1 arc seconds away. These two stars are a true binary, orbiting each other in 246 years, and are at present (2004) at their widest separation; they are at their closest, 0.5 arc seconds, in 2102.

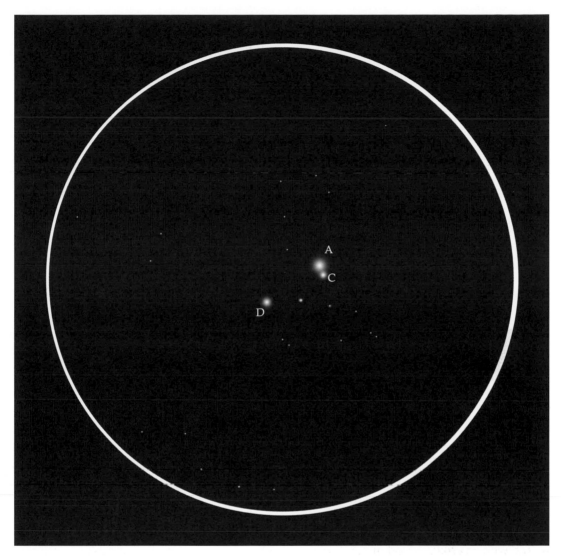

Fig. 12.7 Mizar and Alcor.

Zeta Herculis is a rapidly orbiting binary with a period of only 34.5 years; the stars, of magnitudes 2.8 and 5.7, are at present (2004) only 0.9 arc seconds apart, which is difficult in any telescope. They will be at their widest in 2024, when the separation will be 1.6 arc seconds.

Alpha Herculis, or Rasalgethi, is an attractive yellow and blue double consisting of a magnitude 2.8 primary and a magnitude 5.4 secondary 4.7 arc seconds away. A 100mm (4in) aperture at a medium magnification will give a good view of the pair. This is a binary system with a very long orbital period, 3,600 years.

Nu Draconis is another easy double – in fact one of the easiest in the sky: the two stars, of magnitudes 4.9 and 5.0, are separated by just

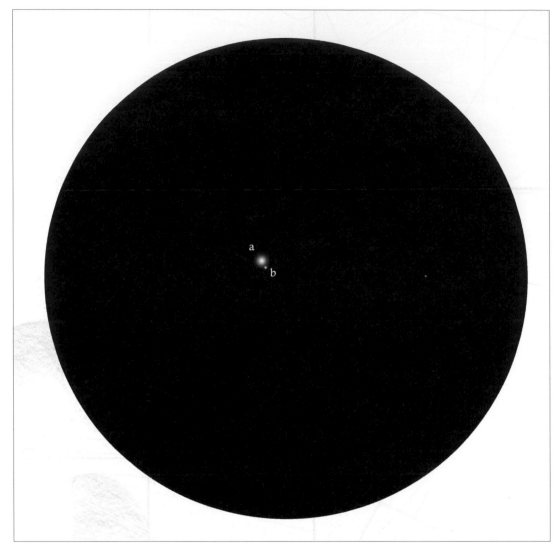

Fig. 12.8 *Epsilon Boötis.*

over a minute of arc, just enough for the unaided eye to resolve. Both stars are white.

Epsilon Lyrae is the famous Double Double, two stars (A and C) separated by 3.3 arc minutes and easily resolved with the naked eye. The stars' magnitudes are 4.7 and 5.1, but through a telescope of at least 100mm (4in) aperture each star is resolved into a true physical double: Epsilon1 (stars A and B) orbit in 1,166 years and are at present (2004) 2.5 arc seconds apart, while Epsilon2 (stars C and D) has an orbital period of 585 years and the stars are at present 2.3 arc seconds apart (Figure 12.9).

Beta Cygni, or Albireo, is undoubtedly one of *the* best if not the best visual binary in the entire northern hemisphere. With a separation of 34 arc seconds, the system is resolvable in a pair of 10 × 50 binoculars, and viewed in even

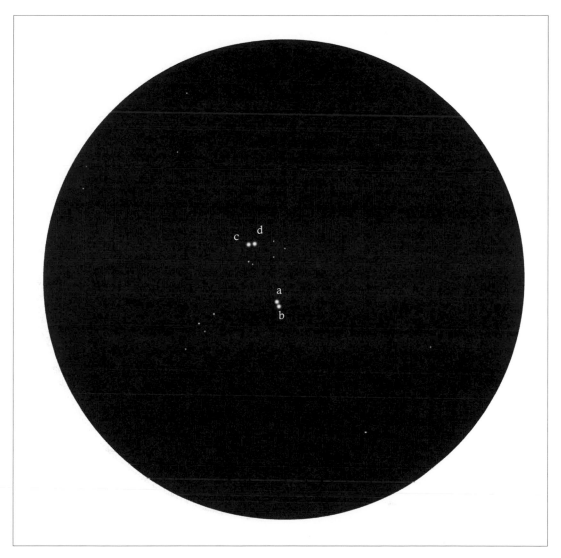

Fig. 12.9 *Epsilon Lyrae.*

the smallest telescope is capable of making people gasp at its beauty: viewed with a magnification of ×50 the sight is never forgotten. Star A is a magnitude 3.4 golden-yellow orb, while its companion is a magnitude 5.4 blue star.

61 Cygni is a wide, easy pair of magnitude 5.4 and 6.1 orange stars separated by 31 arc seconds, visible as a double in 10 × 50 binoculars. This binary system is historically important as in 1838 it became the first star to have its distance measured with any degree of accuracy, by observing its annual parallax against more distant stars; the modern estimate is 11.4 light years. Both stars are red dwarfs and are about 1/40th as luminous as the Sun.

Mu Cygni is a binary system with a highly inclined orbit, period 789 years, and at present

Fig. 12.10 Krüger 60.

(2004) the two stars are only 1.8 arc seconds apart, a tough test for a 150mm (6in) aperture. The separation is closing, and in 2042 will be 1.2 arc seconds. The pair consists of a magnitude 4.5 primary and a magnitude 6.3 secondary.

Krüger 60 is a fascinating binary composed of two dim red dwarfs of magnitudes 9.6 and 11.4 at present (2004) separated by 2.5 arc seconds, making it a tough object for a 150mm (6in) telescope (Figure 12.10). The distance between the two components is about 1,350 million km (850 million miles), comparable to the separation of Saturn and the Sun, and the orbital period is only 44.7 years. The two stars will reach their closest separation in 2014, when they will be 1.4 arc seconds apart.

Southern Double stars, 0° to –90°

Beta Pheonicis is a very nice double of magnitude 3.3 and 4.4 stars with a close 1.4 arc second separation, visible with a 150mm (6in) aperture at high magnification.

Alpha Fornacis is a binary with components of magnitudes 3.8 and 7.2 in a highly elliptical orbit with a period of 314 years. Its present (2004) separation of 5.3 arc seconds makes it an easy double to resolve in any telescope; it will be at its widest in 2063.

Beta Orionis, Rigel, is the bright magnitude 0.2 blue-white star marking the position of the left foot of Orion. It is a tough double

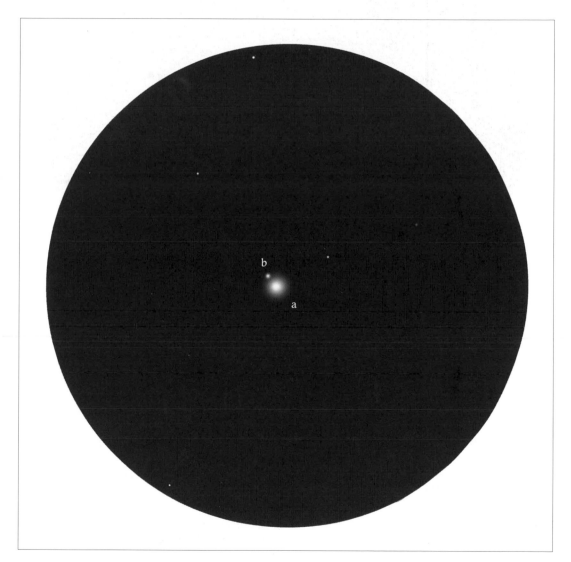

Fig. 12.11 Beta Orionis.

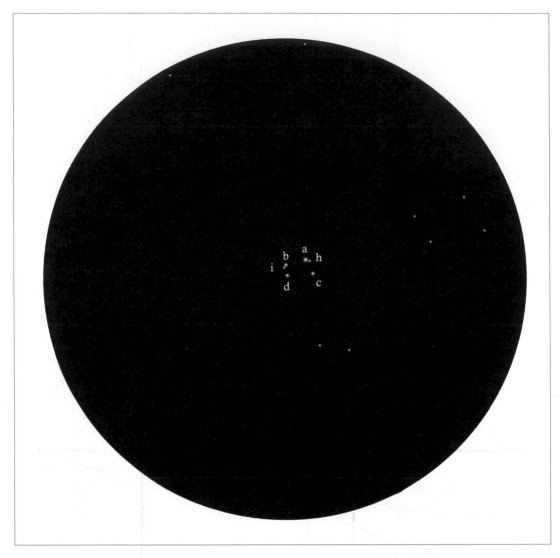

Fig. 12.12 *Theta Orionis.*

because of the large magnitude difference between the primary and its companion, a tiny magnitude 10.4 pinprick of light 9.5 arc seconds away, but this secondary can be seen by careful observation with a 150mm (6in) telescope (Figure 12.11). The actual separation is about 2,450 AU.

Theta Orionis is a magnificent system with a multitude of members, though only the four brightest are readily observable in average seeing conditions. Even a small telescope will show the four main members, of magnitudes 5.1 (A), 6.3 (B), 7.2 (C) and 7.5 (D), and in good seeing conditions a 150mm (6in) telescope may reveal the fainter stars H, (magnitude 11.1, separation 16 arc seconds at 321°) and I (also magnitude 11.1, separation 4.0 arc seconds at PA 122°). Star E is Theta2 Orionis, magnitude 6.4, which lies much farther from the main

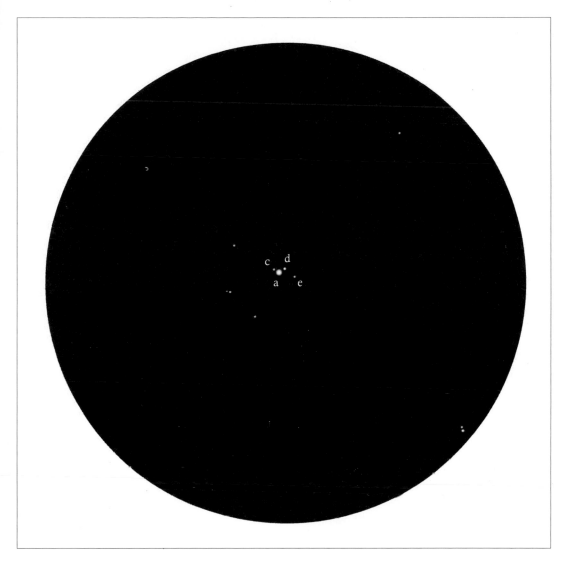

Fig. 12.13 *Sigma Orionis.*

grouping of stars, 135 arc seconds from star A at PA 134° (Figure 12.12). Other members are out of reach of amateur instruments. This system of stars lies in the heart of the Orion Nebula (M42), and is responsible for the nebula's illumination.

Sigma Orionis is a fine quadruple star system visible in a small telescope at high power.

Sigma itself is a magnitude 3.8 white star with a very close companion orbiting in a period of 158 years in a very nearly circular orbit, but at a separation of 0.24 arc seconds it is unlikely to be seen in amateur instruments. Star C is magnitude 10.3 and 11.1 arc seconds distant; star D, magnitude 6.9, is 13.0 arc seconds away; and star E, magnitude 6.9, has a separation of 42 arc seconds (Figure 12.13). The true separation between A and E is at least 20,000 AU, or about a third of a light year. The system is a nice sight in a 100mm (4in) telescope at a medium power.

Beta Monocerotis is an attractive triple star for any size of telescope. The primary star is magnitude 3.8, star B is magnitude 5.0, 7.1 arc seconds away, and star C is magnitude 5.4, 9.9 arc seconds from the primary. A fourth component, much fainter at magnitude 12.2, lies 26 arc seconds away from the main group. All four stars in the group appear brilliant white.

Alpha Canis Majoris, or Sirius, is the brightest star in the entire sky, shining at a brilliant magnitude −1.4, nine times brighter than a 1st-magnitude star. It is also a famous binary, its companion shining at a comparatively feeble magnitude 8.5, 6.8 arc seconds distant. This companion is a white dwarf which orbits Sirius in 50.1 years; the widest separation will be in the year 2022, when the two stars will be 11.3 arc seconds apart. Because of the huge magnitude difference, visual observations are extremely difficult. Under excellent conditions the companion can be seen with a 150mm (6in) aperture at high magnification; an occulting bar would definitely help, although a hexagonal mask to split the diffraction rings may just be enough. Sirius is an astrometric binary. Astronomers noticed a slight oscillation in its motion across the sky; and eventually discovered its companion star telescopically.

Gamma Volantis is an easy wide pair of magnitudes 3.8 and 5.8; 14 arc seconds apart, a nice view in any aperture. The primary is an orange giant.

Gamma Sextantis is a difficult, fast-moving binary with a period of 77.6 years; the magnitudes are 5.1 and 6.1 and the separation is 0.6 arc seconds. The system is at its widest at the present time (2004), the gap will slowly

begin to close until the year 2034, when it will be a mere 0.1 arc seconds – too close for any amateur telescope.

Alpha Crucis is an easy system for a small telescope, consisting of a pair of white stars of magnitudes 0.8 and 1.6; 4.1 arc seconds apart.

Gamma Centauri is a close double of magnitudes 2.2 and 2.9 separated by 0.7 arc seconds. The orbit of these two stars is very elliptical, so they can appear to change position from one year to the next; they will be at their closest in 2017, when they will be a mere 0.1 arc seconds apart, too close for amateur instruments.

Beta Muscae is a slow-moving binary with a period of 383 years in a slightly elliptical orbit. The two stars, of magnitudes 3.0 and 4.0, are at present (2004) 1.3 arc seconds apart, requiring a 150mm (6in) aperture, good observing conditions and a high power to observe. The two are closing; they will reach a 1.3 arc second separation between the years 2018 and 2037.

Gamma Virginis is a magnificent physical double with a 169-year orbit, the stars are of magnitudes 2.7 and 3.6, and the separation is at present (2004) a very difficult 0.4 arc seconds, which is around the closest the two stars ever get. Observers can look forward to watching the two stars slowly begin to separate, though we will have to wait until 2082 for maximum separation, when the stars will be 6.0 arc seconds apart, and a 100mm (4in) aperture should be sufficient to resolve them.

The **Alpha Centauri** system contains our Sun's closest stellar neighbours. The main pair lie at a distance only 4.4 light years, and make a nice double for the telescopic observer. Star A, magnitude 0.1, and its companion, magnitude

1.2, are separated by just over 10 arc seconds at present (2004) and are easily split in a 150mm (6in) aperture, but they will close up, reaching a separation of 4.0 arc seconds in 2015.

Star C, known as Proxima Centauri, is actually the closest star to the Sun. It is a red dwarf of magnitude 12.0 and lies over 2° from its companions, to the SW. The true distance between stars A and C has been calculated to be about 30,000 astronomical units, or about 750 times the distance of Pluto from the Sun.

Mu Lupi is a difficult binary of magnitudes 4.3 and 5.0, a mere 1.1 arc seconds apart, which will need at least a 150mm (6in) aperture and high magnification to split. There is a slightly fainter magnitude 7.0 star 23 arc seconds away at 129° PA from the main group, which is an easier target for smaller telescopes.

Xi Scorpii is a very nice multiple system. Components A and B, magnitudes 4.2 and 5.1, comprise a fast-moving binary system with a period of 45.7 years, and at present (2004) are 0.8 arc seconds apart, which requires 300mm (12in) aperture and high magnification, the separation increasing thereafter. Star C lies 7.6 arc seconds from A and is a magnitude 7.2 orange dwarf. Also in the field of view is star D (which has the separate designation Struve 1999), magnitude 7.4, 280 arc seconds to the south at PA 349°.

Alpha Scorpii, Antares, is the brightest star in Scorpius and is fiery red in colour. Its magnitude

is 1.1 and it has a magnitude 5.4 companion which orbits it in 878 years and is at present (2004) 2.5 arc seconds distant. The orbit of star B as seen from the Earth is edge-on. It is currently (2004) past its widest separation and is slowly moving back towards Antares; by 2146 it will be so close to its parent that it will be unobservable in any telescope.

Gamma Coronae Australis consists of a magnitude 4.2 primary and a magnitude 5.1 secondary, 1.3 arc seconds apart. This is a binary system with an orbital period of 122 years; the two stars were at their closest in 1992, with a separation of 1.3 arc seconds, while maximum separation will be in 2063, when the two stars will be 2.5 arc seconds apart.

Alpha Capricorni is a wide (6 arc minutes) pair of stars visible with the naked eye, of magnitudes 3.6 and 4.3; this is an optical double, the two stars lying at vastly different distances. However, each star is actually a binary. Alpha1 has a very faint magnitude 14.1 companion (component B), 44 arc seconds distant, but the very unequal magnitudes will make spotting it very difficult. Star C is magnitude 9.6, 45 arc seconds away from A at PA 221°.

Zeta Aquarii is a slow-moving binary system with an orbital period of 760 years; the two stars are of magnitudes 3.7 and 4.6 and have a separation of 2.1 arc seconds. They are very slowly widening, but should be visible in a 150mm (6in) telescope in good conditions.

Table of Double Stars

Double stars

The position, magnitudes separations and position angles are given here for all the double stars described in Chapter 12, in the same order as they appear in the text.

Star	RA h m	Dec. ° ′	Magnitudes	Separation ″	PA °
Northern hemisphere					
36 Andromedae	00 54	+23 37	5.5, 6.6	1.0	321
Gamma Arietis	01 53	+19 17	3.9, 4.6	7.6	1
Eta Cassiopeiae	00 49	+57 48	3.5, 7.4	13	319
Gamma Andromedae	02 03	+42 19	2.1, 5.0, 6.3	AB 9.6 BC 0.3	63
Iota Cassiopeiae	02 29	+67 24	4.5, 7.1, 8.5, 8.9	AB 0.1 AC 2.6 CD 207	AB 300 AC 230 CD 58
Alpha Ursae Minoris	02 31	+89 15	2.0, 9.1	18.4	218
7 Tauri	03 34	+24 27	5.9, 6.8	0.8	358
Alpha Tauri	04 35	+16 30	0.9, 13.7	30	110
Alpha Geminorum	07 34	+31 53	1.6, 3.0, 8.8	AB 4.3 AC 72.5	AB 61 AC 164
Zeta Cancri	08 12	+17 38	4.7, 6.1, 6.2	AB 1 AC 5.9	AB 55 AC 70
Gamma Leonis	10 19	+19 50	2.0, 3.6	4.4	125
Iota Leonis	11 23	+10 31	4.0, 6.9	1.9	104
Zeta Ursae Majoris	13 23	+54 55	2.2, 3.9, 3.9	AD 711 AC 14.4	AD 72 AC 152
Epsilon Boötis	14 44	+27 04	2.3, 4.8	2.9	341
Mu Boötis	15 24	+37 22	4.3, 7.2, 7.7	AB 108 DC 2.1	AB 171 BC 5
Zeta Herculis	16 41	+31 36	2.8, 5.7	0.9	0

Alpha Herculis	17 14	+14 23	2.8, 5.4	4.7	104
Nu Draconis	17 32	+55 11	4.9, 5.0	61	312
Epsilon Lyrae	18 44	+39 40	4.7, 6.2, 5.1, 5,5	AB 2.5 AC 207 CD 2.3	AB 349 AC 173 CD 80
Beta Cygni	19 30	+27 57	3.4, 5.4	34	54
61 Cygni	21 06	+38 44	5.4, 6.1	31	150
Mu Cygni	21 44	+28 44	4.5, 6.3	1.8	312
Krüger 60	22 27	+57 41	9.6, 11.4	2.5	67

Southern hemisphere

Beta Pheonicis	01 06	−46 41	3.3, 4.4	1.4	346
Alpha Fornacis	03 12	−28 59	3.8, 7.2	5.3	299
Beta Orionis	05 14	−08 12	0.2, 10.4	9.5	203
Theta Orionis	05 35	−05 23	5.1, 6.3, 7.2, 7.5 6.4 11.1 11.1	AB 12.9 AC 13.1 AD 16.9 AE 135.0 AH 16 AI 4.0	AB 312 AC 56 AD 342 AE 134 AH 321 AI 122
Sigma Orionis	05 38	−02 36	3.8, 5.2, 10.3, 6.9, 6.5	AB 0.24 AC 11.1 AD 13.0 AE 42	AB 101 AC 237 AD 84 AE 61
Beta Monocerotis	06 28	−07 01	3.8, 5.0, 5.4, 12.2	AB 7.1 AC 9.9 AD 26	AB 133 AC 125 AD 56
Alpha Canis Majoris	06 45	−16 43	−1.4, 8.5	6.8	111
Gamma Volantis	07 08	−70 30	3.8, 5.8	14	298
Gamma Sextantis	09 52	−08 06	5.1, 6.1	0.6	55
Alpha Crucis	12 26	−63 05	0.8, 1.6	4.1	114
Gamma Centauri	12 41	−48 57	2.2, 2.9	0.7	341
Beta Muscae	12 46	−68 06	3.0, 4.0	1.3	40
Gamma Virginis	12 41	−01 26	2.7, 3.6	0.4	244
Alpha Centauri	14 39	−60 50	0.1, 1.2, 12.0	AB 10.4 AC 9000	AB 230
Mu Lupi	15 18	−47 52	4.3, 5.0, 7.0	AB 1.1 AC 23	AB 131 AC 129

TABLE OF DOUBLE STARS

Xi Scorpii	16 04	−11 22	4.2, 5.1, 7.2, 7.4	AB 0.8 AC 7.6 AD 280	AB 344 AC 53 AD 349
Alpha Scorpii	16 29	−26 25	1.1, 5.4	2.5	277
Gamma Coronae Australis	19 09	−37 54	4.2, 5.1	1.3	37
Alpha Capricorni	20 17	−12 30	4.3, 14.1, 9.6	AB 44 AC 45	AB 182 AC 221
Zeta Aquarii	22 28	+00 01	3.7, 4.6	2.1	178

Table of Galaxies, Nebulae and Clusters

The position, size, magnitude and type are given here for all the objects described in Chapter 11, in the same order as they appear in the text.

Object	RA h m	Dec. ° ′	Size	Magnitude	Type
Andromeda					
M31	00 43	+41 16	178′ × 63′	4.3	Galaxy
M32	00 43	+40 52	9′ × 7′	8.8	Galaxy
M110	00 40	+41 23	19′ × 11′	8.9	Galaxy
NGC 752	01 58	+37 42	49′	6.6	Open cluster
NGC 891	02 22	+42 20	14′ × 3′	10.8	Galaxy
NGC 7662	23 26	+42 33	17″	9.8	Planetary nebula
Aquarius					
M2	21 33	+00 49	16′	6.6	Globular cluster
M72	20 53	−12 32	7′	9.2	Globular cluster
M73	20 59	−12 38	3′	8.9	Asterism
NGC 7009	21 04	−11 20	29″	8.3	Globular cluster
NGC 7293	22 29	−20	16′	7.6	Planetary nebula

TABLE OF GALAXIES, NEBULAE AND CLUSTERS

Aquila

NGC 6709	18 51	+10 20	13'	7.4	Open cluster

Ara

NGC 6204	16 46	−47 00	5'	8.4	Open cluster

Aries

NGC 772	11 59	+19 00	7' × 4'	11.2	Galaxy

Auriga

M36	05 36	+34 08	12	6.5	Open cluster
M37	05 52	+32 33	23	6.2	Open cluster
M38	05 29	+35 50	21	6.8	Open cluster
IC 2149	05 56	+46 06	8''	11.2	Planetary nebula

Boötes

NGC 5466	14 05	+28 32	9'	9.2	Globular cluster

Camelopardalis

NGC 1502	04 08	+62 21	7'	4.1	Open Cluster
NGC 2403	07 37	+65 35	23' × 12'	8.9	Galaxy

Cancer

M44	08 40	+19 59	95'	3.9	Open cluster
M67	08 50	+11 49	29'	7.3	Open cluster

Canes Venatici

M3	13 42	+28 23	18'	6.3	Globular cluster
M51	13 30	+47 12	11' × 8'	8.9	Galaxy
M63	13 16	+42 02	13' × 8'	9.3	Galaxy

M94	12 50	+41 07	12′ × 11′	8.8	Galaxy
M106	12 19	+47 18	17′ × 6′	9.1	Galaxy
NGC 4449	12 28	+44 05	6′ × 4′	10.1	Galaxy

Canis Major

| M41 | 06 47 | −20 44 | 38′ | 5.0 | Open cluster |
| NGC 2362 | 07 19 | −24 56 | 8′ | 3.8 | Open cluster |

Capricornus

| M30 | 21 40 | −23 11 | 12′ | 6.9 | Globular cluster |

Cassiopeia

M52	23 24	+61 35	12′	8.2	Open cluster
M103	01 33	+60 42	6′	6.9	Open cluster
NGC 129	00 30	+60 13	21′	9.8	Open cluster
NGC 281	00 53	+56 38	4′	7.4	Diffuse nebulae
NGC 457	01 19	+58 19	13′	5.1	Open cluster

Centaurus

| NGC 5128 | 13 25 | −43 01 | 29′ × 21′ | 7.7 | Galaxy |
| NGC 5139 | 13 26 | −47 28 | 55′ | 3.9 | Globular cluster |

Cepheus

| NGC 40 | 00 13 | +72 33 | 48″ | 10.7 | Planetary nebula |
| NGC 188 | 00 44 | +85 22 | 13′ | 9.3 | Open cluster |

Cetus

| M77 | 02 43 | −00 01 | 7′ × 6′ | 9.5 | Galaxy |
| NGC 246 | 00 47 | −11 52 | 4′ | 8.0 | Planetary nebula |

TABLE OF GALAXIES, NEBULAE AND CLUSTERS

NGC 247	00 47	−20 44	21′ × 6′	9.7	Galaxy

Columba

NGC 1851	05 14	−40 02	12′	7.1	Globular Cluster

Coma Berenices

M53	13 13	+18 10	13′	7.7	Globular cluster
M64	12 57	+21 41	11′ × 5′	9.3	Galaxy
M85	12 25	+18 11	7′ × 5′	9.2	Galaxy
M88	12 32	+14 25	7′ × 4′	10.3	Galaxy
M91	12 35	+14 30	5′ × 4′	10.9	Galaxy
M98	12 14	+14 24	10′ × 3′	10.9	Galaxy
M99	12 19	+14 55	5′	10.4	Galaxy
M100	12 23	+15 49	7′ × 6′	10.1	Galaxy
NGC 4559	12 36	+27 57	12′ × 4′	10.5	Galaxy
NGC 4565	12 36	+25 59	15′ × 2′	10.6	Galaxy

Corvus

NGC 4038/39	12 02	−18 52	6′ × 4′	10.9	Galaxy
NGC 4361	12 24	−18 47	1′	10.3	Planetary nebula

Crux

NGC 4755	12 54	−60 19	10′	5.2	Open cluster
Coalsack	12 30	−63 48	7 × 4°	—	Dark nebula

Cygnus

M29	20 24	+38 32	6'	7.5	Open cluster
M39	21 32	+48 26	31'	5.3	Open cluster
NGC 6826	19 44	+50 32	25''	9.8	Planetary nebula
NGC 6910	20 23	+40 46	7'	7.3	Open cluster
NGC 6946	20 34	+60 10	11' × 10'	9.7	Galaxy
NGC 6960	20 45	+30 43	200'	7.0	Supernova remnant
NGC 6992	20 56	+31 43	60'	7.0	Supernova remnant
NGC 6995	20 57	+31 13	12'	7.0	Supernova remnant
NGC 7000	21 01	+44 12	120'	4.0	Diffuse nebula

Delphinus

NGC 6934	20 34	+07 25	7'	8.9	Globular cluster
NGC 6891	20 15	+12 42	15''	11.7	Planetary nebula
NGC 7006	21 01	+16 12	4'	10.6	Globular cluster

Dorado

NGC 2070	05 38	−69 09	40' × 25'	8.3	Diffuse nebula
LMC	05 23	−69 43	11° × 9°	0.9	Galaxy

Draco

NGC 5907	15 15	+56 19	12' × 1'	11.1	Galaxy
NGC 6543	17 58	+66 37	20''	8.8	Planetary nebula

TABLE OF GALAXIES, NEBULAE AND CLUSTERS

Eridanus

NGC 1535	04 14	−12 43	21″	9.6	Planetary nebula

Fornax

NGC 1316	03 22	−37 12	11′ × 8′	9.8	Galaxy
NGC 1360	03 33	−25 52	6′	9.6	Planetary nebula

Gemini

M35	06 09	+24 20	28′	5.6	Open cluster
NGC 2158	06 07	+24 05	5′	12.1	Open cluster
NGC 2371/2	07 25	+29 29	44″	13.0	Planetary nebula
NGC 2392	07 29	+20 54	20″	9.9	Planetary nebula

Hercules

M13	16 42	+36 28	20′	5.9	Globular cluster
M92	17 17	+43 08	14′	6.5	Globular cluster
NGC 6229	16 46	+47 31	4.5′	9.4	Globular cluster

Horologium

NGC 1512	04 03	−34 20	9′ × 5′	11.1	Galaxy

Hydra

M48	08 14	−05 48	54′	5.5	Open cluster
M68	12 39	−26 45	11′	8.2	Globular cluster
M83	13 37	−29 52	13′ × 12′	7.9	Galaxy
NGC 3242	10 25	+18 38	25″	8.6	Planetary nebula

Lacerta

NGC 7209	22 05	+46 29	24'	7.8	Open cluster

Leo

M65	11 19	+13 05	10' × 2'	10.2	Galaxy
M66	11 20	+12 59	9' × 4'	9.6	Galaxy
M95	10 44	+11 42	7' × 5'	10.5	Galaxy
M96	10 47	+11 49	7' × 5'	10.1	Galaxy
M105	10 48	+12 35	5' × 4'	10.2	Galaxy
NGC 2903	09 32	+21 30	13' × 5'	9.6	Galaxy

Lepus

M79	05 25	−24 31	9.6'	7.7	Globular cluster
IC 418	05 27	−12 41	12''	10.7	Planetary nebula

Libra

NGC 5897	15 17	−21 00	11'	8.4	Globular cluster
Merrill 2-1	15 22	−23 37	6''	11.6	Planetary nebula

Lupus

NGC 5824	15 04	−33 05	7'	9.1	Globular cluster

Lynx

NGC 2419	07 38	+38 52	5'	10.3	Globular cluster

Lyra

M56	19 17	+30 11	7'	8.2	Globular cluster

TABLE OF GALAXIES, NEBULAE AND CLUSTERS

M57	18 54	+33 02	1'	9.7	Planetary nebula
NGC 6791	19 20	+37 51	15'	9.5	Open cluster

Monoceros

M50	07 03	−08 20	16'	7.2	Open cluster
NGC 2237	06 30	+05 03	80'	5.5	Diffuse nebula
NGC 2261	06 39	+08 45	2'	var.	Reflection nebula
NGC 2264	06 41	+09 53	20'	4.1	Open cluster
NGC 2301	06 52	+00 28	12'	6.3	Open cluster

Ophiuchus

M9	17 19	−18 31	12'	7.8	Globular cluster
M10	16 57	−04 06	20'	6.6	Globular cluster
M12	16 47	−01 57	16'	6.1	Globular cluster
M14	17 37	−03 15	12'	7.6	Globular cluster
M19	17 02	−26 16	17'	6.8	Globular cluster
M62	17 01	−30 07	13'	6.4	Globular cluster
M107	16 32	−13 03	13'	7.8	Globular cluster

Orion

M42	05 35	−05 27	66' × 60'	4.0	Diffuse nebula
M43	05 36	−05 16	20' × 15'	9.0	Diffuse nebula
M78	05 47	+00 05	8' × 6'	8.0	Reflection nebula
NGC 1662	04 48	+10 56	20'	8.0	Open cluster
NGC 2169	06 08	+13 57	6'	7.0	Open cluster

NGC 1981	05 35	−04 24	24'	4.2	Open cluster
NGC 2022	05 42	+09 05	19''	12.4	Planetary nebula

Pegasus

M15	21 30	+12 10	18'	6.4	Globular cluster

Perseus

M34	02 42	+42 47	35'	5.8	Open cluster
M76	01 42	+51 34	1' × 1'	12.2	Planetary nebula
NGC 869	02 19	+57 08	29'	4.3	Open cluster
NGC 884	02 22	+57 06	29'	4.3	Open cluster
NGC 1499	04 03	+36 23	150'	5.0	Diffuse nebula

Pisces

M74	01 37	+15 47	10' × 9'	10.0	Galaxy

Puppis

M46	07 42	−14 49	27'	6.6	Open cluster
M47	07 36	−14 28	29'	4.3	Open cluster
M93	07 45	−23 52	22'	6.5	Open cluster
NGC 2423	07 37	−13 51	19'	7.0	Open cluster
NGC 2451	07 45	−37 57	45'	3.7	Open cluster
NGC 2477	07 52	−38 51	27'	5.7	Open cluster
NGC 2546	08 12	−37 37	40'	5.2	Open cluster

TABLE OF GALAXIES, NEBULAE AND CLUSTERS

Pyxis

NGC 2627	08 37	−29 56	11′	8.4	Open cluster
NGC 2818	09 16	−36 36	50″	13.0	Open cluster

Sagitta

M71	19 54	+18 47	4′	8.3	Globular cluster

Sagittarius

M8	18 04	−24 23	90′ × 40′	5.0	Diffuse nebula
M17	18 21	−16 11	46′ × 37′	6.0	Diffuse nebula
M18	18 20	−17 08	9′	7.5	Open cluster
M20	18 03	−23 02	29′ × 27′	6.3	Diffuse nebula
M21	18 05	−22 30	13′	7.2	Open cluster
M22	18 37	−23 54	32′	5.1	Globular cluster
M23	17 57	−19 01	27′	5.5	Open cluster
M25	18 32	−19 15	32′	6.2	Open cluster
M28	18 24	−24 52	14′	6.9	Globular cluster
M54	18 55	−20 39	12′	7.7	Globular cluster
M55	19 40	−30 58	19′	6.3	Globular cluster
M69	18 31	−32 31	19′	7.7	Globular cluster
M70	18 43	−32 18	8′	7.8	Globular cluster
M75	20 06	−21 55	6.8′	8.6	Globular cluster

Scorpius

M4	16 23	−26 32	36'	5.4	Globular cluster
M6	17 40	−32 13	15'	4.6	Open cluster
M7	17 54	−34 49	80'	3.3	Open cluster
M80	16 17	−22 59	10'	7.2	Globular cluster
NGC 6124	16 25	−40 38	29'	6.3	Globular cluster
NGC 6231	16 54	−41 46	14'	3.4	Open cluster
NGC 6242	16 55	−39 28	9'	8.2	Open cluster

Sculptor

NGC 55	00 15	−39 13	27' × 5'	9.6	Galaxy
NGC 253	00 47	−25 17	27' × 6'	8.2	Galaxy
NGC 288	00 52	−26 35	13'	8.1	Globular cluster
NGC 300	00 54	−37 40	19' × 13'	8.8	Galaxy
NGC 613	01 34	−29 25	5' × 3'	11.0	Galaxy
NGC 7507	23 12	−28 32	3' × 3'	11.6	Galaxy

Scutum

M11	18 51	−06 16	14'	6.1	Open cluster
M26	18 45	−09 24	14'	9.0	Open cluster

TABLE OF GALAXIES, NEBULAE AND CLUSTERS

Serpens

M5	15 19	+02 05	23'	5.7	Globular cluster
M16	18 19	−13 47	6'	6.0	Open cluster
Palomar 5	15 16	−00 06	3'	11.8	Globular cluster

Sextans

NGC 3115	10 05	−07 44	7' × 3'	10.1	Galaxy

Taurus

M1	05 34	+22 01	6' × 4'	8.4	Supernova remnant
M45	03 47	+24 07	110'	1.2	Open cluster
Hyades	04 26	+15 51	300'	0.8	Open cluster
NGC 1514	04 09	+30 47	2'	10.0	Planetary nebula
NGC 1647	16 47	+19 04	45'	6.2	Open cluster
NGC1746	05 03	+23 49	41'	6.1	Open cluster

Traiangulum

M33	01 34	+30 39	62' × 39'	5.7	Galaxy
NGC 672	01 47	+27 25	6' × 2'	11.6	Galaxy
NGC 925	02 27	+33 34	11' × 6'	10.9	Galaxy

Tucana

NGC 104	00 24	−72 04	50'	4.0	Globular cluster
NGC 292	00 52	−72 50	5.3° × 3.4°	2.8	Galaxy

Ursa Major

M40	12 22	+58 05	—	8.0	Double star
M81	09 56	+69 04	26' × 14'	7.8	Galaxy

M82	09 56	+69 41	11′ × 5′	9.2	Galaxy
M97	11 15	+55 01	3′	12.0	Planetary nebula
M101	14 03	+54 21	27′ × 26′	7.7	Galaxy
M108	11 11	+55 40	8′ × 2′	10.9	Galaxy
M109	11 58	+53 23	8′ × 5′	10.8	Galaxy
NGC 3675	11 26	+43 35	6′ × 3′	11.0	Galaxy
NGC 3726	11 33	+47 01	6′ × 4′	10.9	Galaxy
NGC 3877	11 46	+47 29	5′ × 1′	11.8	Galaxy
NGC 4026	11 59	+50 57	5′ × 1′	11.7	Galaxy

Vela

NGC 2547	08 10	−49 16	20′	5.0	Galaxy
NGC 3132	10 07	−40 26	30″	8.2	Planetary nebula
NGC 3201	10 07	−46 24	20′	6.9	Globular cluster

Virgo

M84	12 25	+12 53	6′ × 4′	10.1	Galaxy
M86	12 26	+12 57	10′ × 7′	9.3	Galaxy
M87	12 31	+12 24	8′	9.5	Galaxy
M104	12 39	−11 37	8′ × 4′	9.1	Galaxy
NGC4378	12 25	+04 55	3′ × 2′	12.3	Galaxy

Vulpecula

M27	19 59	+22 43	8′ × 4′	7.6	Planetary nebula
NGC 6940	20 34	+28 18	31′	7.2	Open cluster
Collinder 399	19 25	+20 11	60′	4.8	Open cluster

Glossary

absolute magnitude
The apparent magnitude that a star would possess it if were 10 parsecs (32.6 light years) from Earth. In this way, absolute magnitude provides a direct comparison of the brightness of stars.

achromat
A two-element telescope lens or eyepiece designed to reduce the false colours introduced into images by chromatic aberration.

Airy disk
The central spot of the image of a star formed in a telescope; it is surrounded by diffraction rings.

altazimuth mounting
A mounting that allows a telescope to rotate about two axes: in azimuth (horizontally), and in altitude (vertically). This is a common mounting, though tracking objects is difficult because the telescope needs to be moved about both axes continually (*compare* equatorial mounting). Dobsonian telescopes are usually mounted this way.

Antoniadi scale
A scale of 1 to 5 indicating the quality of seeing – the steadiness of the atmosphere. It goes from 1, representing perfect seeing, to 5, which represents the worst seeing conditions.

aperture
The diameter of an objective lens or primary mirror usually expressed in millimetres (or inches). The larger the aperture, the greater the light-gathering ability of the telescope (or binoculars).

apochromat
A lens or optical instrument designed to provide a high degree of correction for the false colours caused by chromatic aberration. Apochromatic eyepieces usually have three elements.

apparent field of view
The field of view an eyepiece will give at a certain magnification.

apparent magnitude
The brightness of a celestial object indicated on a numerical scale on which the brightest star (Sirius) has magnitude −1.4, and the faintest star visible to the unaided eye has magnitude 6.5. A decrease of one unit represents an increase in apparent brightness by a factor of 2.512.

arc minute (')
A measure of angle, equal to 1/60 of a degree.

arc second (2')
A measure of angle, equal to 1/60 of an arc minute (1/3,600th of a degree).

asterism
A distinctive pattern formed by a group of stars within a constellation.

astigmatism
A defect in a lens or mirror (or the human eye) that causes light rays which are off the central axis of the lens or mirror to form an ellipse or straight line instead of being brought to a point focus.

astronomical unit (AU)
The mean distance from the Earth to the Sun, approximately 149.6 million km (93 million miles).

averted vision
A technique for detecting faint objects in which the observer looking slightly to one side of the object being studied.

Barlow lens
An extra lens used in conjunction with a telescope's eyepiece to increase magnification, usually by a factor of two or three, by increasing the effective focal length of the telescope.

binary star
Two stars bound by their mutual gravity and orbiting a common centre of mass. Binary stars are twins in the sense that they formed together out of the same interstellar cloud.

Cassegrain telescope
A reflecting telescope consisting of a primary mirror with a central hole through which the light is reflected from a secondary mirror to a focus (the Cassegrain focus) behind the primary mirror. The design, in the form of the Schmidt–Cassegrain telescope, makes for compact and portable telescopes.

catadioptric
A telescope consisting of a primary mirror with a full-aperture lens, mounted ahead of it, to correct for spherical aberration in the primary mirror.

celestial equator
The projection of the Earth's equator onto the celestial sphere.

celestial poles
The projections of the Earth's poles onto the celestial sphere.

celestial sphere
The projection the night sky onto an imaginary hollow sphere of infinite radius surrounding the Earth and centred on the observer. It is the basis of sky charts, and of celestial coordinates (*see* declination, right ascension).

chromatic aberration
A defect in a non-achromatic lens in which different wavelengths of light are refracted by slightly different amounts and are thus brought to a focus at slightly different distances from the lens.

coma
A defect in an optical system that gives rise to a blurred, pear-shaped, comet-like image.

dark adaptation
Increasing one's visual sensitivity in darkness by keeping the eyes shielded from bright lights.

declination (dec.)
An astronomical coordinate in which the distance of an object on the celestial sphere is expressed as an angular distance north or south of the celestial equator; it is the equivalent of terrestrial latitude. Declination is positive for objects north of the celestial equator, and negative for objects south. *Compare* right ascension.

diffraction
The spreading out of light as it passes over a sharp edge, or through a narrow slit (narrow in relation to the wavelength of light). The obstructions in the tube of a telescope can give

rise to diffraction, reducing the contrast of the image and causing rings or spikes to appear around stellar images (*see also* Airy disk).

Dobsonian telescope
A Newtonian telescope mounted on a form of altazimuth mount, in which the mirror end of the telescope sits in a rotatable 'rocker box'. This design allows large-aperture telescopes to be made relatively cheaply.

doublet
A lens system consisting two elements, used to reduce chromatic aberration.

dwarf
An 'ordinary' star, like the Sun, which has yet to evolve into a giant or supergiant star.

equatorial mounting
A telescope mounting in which one axis (the polar axis) is aligned with the observer's visible celestial pole, and the other axis (the declination axis), to which the telescope is fixed, is at right angles to the polar axis. Once an object is centred in the telescope's field of view, only the polar axis needs to be adjusted to keep the object in view.

exit pupil
The minimum diameter of the beam of light that leaves an eyepiece. It is equal to the aperture divided by the magnification.

extended object
A celestial object, such as a nebula, which presents a finite surface area (as opposed to a stellar point) to the observer.

eye lens
The lens in an eyepiece that is closest to the eye.

eyepiece (or ocular)
A system of lenses used to magnify the image formed at the focus of a telescope or binoculars. The magnification that an eyepiece provides is its focal length divided into that of the telescope.

eye relief
The distance the eye must be positioned behind the eye lens to see the full field of view.

field of view
The area that is visible through an optical instrument. *See also* apparent field of view.

field star
Any star that is visible in an eyepiece view, apart from the object being specifically observed.

finder (finderscope)
A small, low-power, wide-field telescope attached to the main telescope to help in locating celestial objects.

focal length
The distance from a lens or mirror to its focus.

focal ratio
The focal length of a telescope divided by its aperture.

focuser
The mechanism that holds the eyepiece and allows it to be moved in and out to focus the image.

galaxy
A star system containing millions to many billions of stars, along with dust and gas, held together by gravity. There are three main classes of galaxy: elliptical, spiral and barred, named after their appearance.

globular cluster
A spherical cluster of older stars. Globular clusters are often found in the halo of a galaxy, away from the galactic plane.

hydrogen-beta
A weak emission line located at 486nm in the blue-green region of the spectrum.

integrated magnitude
The magnitude an extended object would have if all of its light were concentrated at a star-like point.

light bucket
A slang term for a large-aperture telescope, such as big Dobsonian telescopes.

light year
A unit of distance equal to the distance travelled by light in one year, equal to 9.46 million million km (5.88 million million miles).

limiting magnitude
(1) The magnitude of the faintest object that can just be detected by a telescope. (2) The magnitude of the faintest stars plotted in a star atlas.

luminosity
The total radiation emitted into space by a celestial object such as a star.

magnification
The apparent increase in angular size of a celestial object given by a particular telescope–eyepiece combination. The magnification can be calculated by dividing the focal length of the telescope by the focal length of the eyepiece.

magnitude
See absolute magnitude, apparent magnitude.

Messier object
A deep-sky object in the list drawn up by Charles Messier.

nebula
A cloud of gas or dust in space. See also planetary nebula.

Newtonian telescope
A type of reflecting telescope with a parabolic primary mirror; the most frequently used type of telescope for deep-sky observing.

objective
The main light-gathering element of a telescope or binoculars. In a refractor, this is the large lens at the front, sometimes called an object glass. In a reflector, it is a mirror.

occulting bar
A small, thin opaque piece of material fixed to the field stop of an eyepiece to block the light from a bright object to make a faint object easier to observe.

ocular
See eyepiece.

open cluster
A group of gravitationally bound stars in the spiral arm of a galaxy.

optical double
A pair of stars whose apparent closeness when viewed through a telescope is a line of sight effect, the stars actually being separated by a great distance.

orbit
The path of one body around another under the influence of gravity.

physical double
A double star in which, unlike an optical double, the two components are physically associated – a binary star.

planetary nebula
An expanding shell of gas ejected by a star late in its life.

position angle
The direction of one celestial body with respect

GLOSSARY

to another on the celestial sphere. It is measured in degrees, from north (0°) through east (90°), south (180°) and west (270°).

primary mirror
The principal, light-gathering mirror in a reflecting telescope.

reflecting telescope (reflector)
A telescope in which the main light-gathering element is a mirror. The most common type for amateur astronomy is the Newtonian; other types are the Cassegrain and the Maksutov.

refracting telescope (refractor)
A telescope in which the main light-gathering element is a lens.

red giant
A star nearing the end of its life which has swelled to around 25 times the Sun's diameter and has a low surface temperature, e.g. Arcturus in Boötes.

resolution
The level of detail visible in a telescopic image; a low-resolution image shows only large-scale features, a high-resolution one shows many small details.

right ascension (RA)
An astronomical coordinate in which the distance of an object on the celestial sphere is measured eastwards, from the vernal equinox, the point where the Sun crosses the celestial equator in the spring; it is the equivalent of terrestrial longitude. It is measured in hours, from 0 to 24; 1 hour of RA is equivalent to 15 degrees. Compare declination.

Schmidt–Cassegrain telescope (SCT)
A catadioptric telescope consisting of a spherical primary mirror, a secondary mirror and a corrector plate.

secondary mirror
A mirror in a reflecting telescope positioned at a 45° angle which serves to reflect light from the primary mirror to the eyepiece.

seeing
The stability of an image formed in a telescope as affected by turbulence in the atmosphere. It is measured on the Antoniadi scale.

separation
The angular distance between two, usually close celestial objects.

star cluster
See globular cluster, open cluster.

supernova
A cataclysmic explosion that marks the end of a massive star's life; a supernova can briefly outshine its host galaxy.

surface brightness
The brightness of an extended object, such as a galaxy, expressed as its magnitude divided by its area (reckoned usually in square arc minutes).

transparency
A measure of the clarity of the atmosphere, normally indicated by the magnitude of the faintest naked-eye star visible close to the zenith.

visual binary
A binary star whose components are far enough apart to be seen separately in a telescope.

white dwarf
A dim, whitish star of up to 1.4 solar masses and about the size of the Earth, with a very high density.

zenith
The point on the celestial sphere directly above the observer.

Bibliography

Guidebooks

Burnham's Celestial Handbook, Robert Burnham, Jr (3 vols, Dover, 1978)

David Levy's Guide to the Night Sky, David Levy (Cambridge University Press, 2001)

Deep Sky Wonders, Walter Scott Houston (Sky Publishing, 2001)

Messier's Nebulae and Star Clusters, Kenneth Glyn Jones (Cambridge University Press, 1991)

The Monthly Sky Guide, Ian Ridpath and Wil Tirion (Cambridge University Press, 2003)

The Night Sky Observers Guide, George Robert Kepple and Glen W. Sanner (Willmann-Bell, 1999)

Nightwatch, Terence Dickinson (Firefly, 1996)

Star-Hopping for Backyard Astronomers, Alan M. MacRobert (Sky Publishing, 1993)

Star-Hopping: Your Vista to Viewing the Universe, Robert Garfinkle (Cambridge University Press, 1994)

Skywatching, David Levy (Nature/Time-Life, 1994)

365 Starry Nights, Leo Enright (Royal Astronomical Society of Canada, 1999)

Turn Left at Orion, Guy J. Consolmagno and Dan M. Davis (Cambridge University Press, 2001)

Binoculars and Telescopes

Binocular Astronomy, Craig Crossen and Wil Tirion (Willmann-Bell, 1992)

Choosing and Using a Schmidt–Cassegrain Telescope, Rod Mollise (Springer, 2001)

Exploring the Night Sky with Binoculars, Patrick Moore (4th edn, Cambridge University Press, 2000)

How to Use an Astronomical Telescope, James Muirden (reprint, Fireside, 1988)

Using the Meade ETX, Mike Weasner (Springer, 2002)

Star Ware, Philip S. Harrington (Wiley, 2002)

The 20-cm Schmidt–Cassegrain Telescope, Peter L. Manly (Cambridge University Press, 2000)

Atlases and Catalogues

Bright Star Atlas, Wil Tirion (Willmann-Bell, 1990)

Cambridge Star Atlas 2000.0, Wil Tirion (3rd edn, Cambridge University Press, 2001)

The Millennium Star Atlas, Roger W. Sinnott and Michael A.C. Perryman (Sky Publishing, 1997)

NGC 2000.0, Roger Sinnott (ed.) (Cambridge University Press, 1989)

Norton's 2000.0, Ian Ridpath (ed.) (20th edn, Pi Press, 2004). For addenda and corrigenda *see* **www.ianridpath.com/books/ nortonpage2.htm**

The Observer's Sky Atlas, Erich Karkoschka (Springer, 1998)

Observing Handbook and Catalogue of Deep-Sky Objects, Chris Luginbuhl and Brian Skiff (Cambridge University Press, 1989)

Sky Atlas 2000.0, Wil Tirion (Sky Publishing/Cambridge University Press, 1981)

Sky Atlas 2000.0 Companion, Robert A.

BIBLIOGRAPHY

Strong and Roger Sinnott (2nd edn, Sky Publishing/Cambridge University Press, 2000)

Uranometria 2000.0, Wil Tirion, Barry Rappaport and George Lovi (Willmann-Bell, 2001)

The Deep-Sky Field Guide to Uranometria 2000.0, Murray Cragin, James Lucyk and Barry Rappaport (2nd edn, Willmann-Bell, 2001)

Software

Cartes du Ciel 15MB file downloadable from **www.stargazing.net/astropc**

MegaStar Willmann-Bell, **www.willbell.com/software/megastar/index.htm**

Redshift Focus Multimedia Ltd, **www.focusmm.co.uk/store/productpages/productinfo/ffb032.html**

SkyMap Pro SkyMap Software, **www.skymap.com**

SkyTools Capella Soft, **www.skyhound.com/skytools.html**

Starry Night Imaginova Corp., **www.starrynight.com/products.html**

The Sky Software Bisque, **www.bisque.com/Products/TheSky6**

Online Deep-Sky Catalogues

'Astronomical League's Deep Sky Binocular Club' The Astronomical League, **seds.lpl.arizona.edu/messier/xtra/similar/albi-d.html**

'Deepsky Observer's Companion' Astronomical Society of Southern Africa, **www.saao.ac.za/assa/html/doc_home_page.html** for the southern sky.

'The Interactive NGC Catalog Online' **www.seds.org/~spider/ngc/ngc.html**

'List of Common Deep Sky Catalogs', a site that links to many on-line catalogues: **www.seds.org/messier/xtra/supp/cats.html**

'The RASC's Deep Sky Challenge Objects list' Royal Astronomical Society of Canada, **seds.lpl.arizona.edu/messier/xtra/similar/rasc-dsc.html**

'Deep Sky Browser', an easy-to-use front-end for the Deep Sky Database which includes data and information on about 500,000 deep-sky objects – Mikkel Steine, **messier45.com/cgi-bin/dsdb/dsb.pl**

Index

INDEX

INDEX